Multi-Objective Decision Making

Synthesis Lectures on Artificial Intelligence and Machine Learning

Editors
Ronald J. Brachman, *Yahoo! Labs*
Peter Stone, *University of Texas at Austin*

Essentials of Game Theory: A Concise Multidisciplinary Introduction
Kevin Leyton-Brown and Yoav Shoham
2008

A Concise Introduction to Multiagent Systems and Distributed Artificial Intelligence
Nikos Vlassis
2007

Intelligent Autonomous Robotics: A Robot Soccer Case Study
Peter Stone
2007

Multi-Objective Decision Making

Diederik M. Roijers and Shimon Whiteson

ISBN: 978-3-031-00448-3   paperback
ISBN: 978-3-031-01576-2   ebook

DOI 10.1007/978-3-031-01576-2

A Publication in the Springer series
SYNTHESIS LECTURES ON ARTIFICIAL INTELLIGENCE AND MACHINE LEARNING

Lecture #34
Series Editors: Ronald J. Brachman, *Yahoo! Labs*
                Peter Stone, *University of Texas at Austin*
Series ISSN
Print 1939-4608   Electronic 1939-4616

Multi-Objective Decision Making

Diederik M. Roijers
University of Oxford
Vrije Universiteit Brussel

Shimon Whiteson
University of Oxford

SYNTHESIS LECTURES ON ARTIFICIAL INTELLIGENCE AND MACHINE LEARNING #34

ABSTRACT

Many real-world decision problems have multiple objectives. For example, when choosing a medical treatment plan, we want to maximize the efficacy of the treatment, but also minimize the side effects. These objectives typically conflict, e.g., we can often increase the efficacy of the treatment, but at the cost of more severe side effects. In this book, we outline how to deal with multiple objectives in decision-theoretic planning and reinforcement learning algorithms. To illustrate this, we employ the popular problem classes of *multi-objective Markov decision processes* (MOMDPs) and *multi-objective coordination graphs* (MO-CoGs).

First, we discuss different use cases for multi-objective decision making, and why they often necessitate explicitly multi-objective algorithms. We advocate a *utility-based* approach to multi-objective decision making, i.e., that what constitutes an optimal solution to a multi-objective decision problem should be derived from the available information about user utility. We show how different assumptions about user utility and what types of policies are allowed lead to different solution concepts, which we outline in a taxonomy of multi-objective decision problems.

Second, we show how to create new methods for multi-objective decision making using existing single-objective methods as a basis. Focusing on planning, we describe two ways to creating multi-objective algorithms: in the inner loop approach, the inner workings of a single-objective method are adapted to work with multi-objective solution concepts; in the outer loop approach, a wrapper is created around a single-objective method that solves the multi-objective problem as a series of single-objective problems. After discussing the creation of such methods for the planning setting, we discuss how these approaches apply to the learning setting.

Next, we discuss three promising application domains for multi-objective decision making algorithms: energy, health, and infrastructure and transportation. Finally, we conclude by outlining important open problems and promising future directions.

KEYWORDS

artificial intelligence, decision theory, decision support systems, probabilistic planning, multi-agent systems, multi-objective optimization, machine learning

Contents

Preface

Many real-world decision problems have multiple, possibly conflicting, objectives. For example, an autonomous vehicle typically wants to minimize both travel time and fuel costs, while maximizing safety; when seeking medical treatment, we want to maximize the probability of being cured, but minimize the severity of the side-effects, etcetera.

Although interest in multi-objective decision making has grown in recent years, the majority of decision-theoretic research still assumes only a single objective. In this book, we argue that multi-objective methods are underrepresented and present three scenarios to justify the need for explicitly multi-objective approaches. Key to these scenarios is that, although the utility the user derives from a policy—which is what we ultimately aim to optimize—is scalar, it is sometimes impossible, undesirable, or infeasible to formulate the problem as single-objective at the moment when the policies need to be planned or learned. We also present the case for a *utility-based* view of multi-objective decision making, i.e., that the appropriate multi-objective solution concept should be derived from what we know about the user's utility function.

This book is based on our research activities over the years. In particular, the survey we wrote together with Peter Vamplew and Richard Dazeley [Roijers et al., 2013a] forms the basis of how we organize concepts in multi-objective decision making. Furthermore, we use insights from our work on multi-objective planning over the years, particularly in the context of the PhD research of the first author [Roijers, 2016]. Another important source for writing this book were the lectures we gave on the topic at the University of Amsterdam, and the tutorials we did at the IJCAI-2015 and ICAPS-2016 conferences, as well as the EASSS-2016 summer school.

Aim and Readership   This book aims to provide a structured introduction to the field of multi-objective decision making, and to make the differences with single-objective decision theory clear. We hope that, after reading this book, the reader will be equipped to conduct research in multi-objective decision-theory or apply multi-objective methods in practice.

We expect our readers to have a basic understanding of decision theory, at a graduate or undergraduate level. In order to remain accessible to a wide range of readers, we provide intuitive explanations and examples of key concepts before formalizing them. In some cases, we omit detailed proofs of theorems in order to better focus on the intuition behind and implications of these theorems. In such cases, we provide references to the detailed proofs.

Outline   This book is structured as follows. In Chapter 1, we motivate multi-objective decision making by providing examples of multi-objective decision problems and scenarios that require explicitly multi-objective solution methods. In Chapter 2, we introduce two popular classes of decision problems that we use throughout the book to illustrate specific algorithms and general theoretical results. In Chapter 3, we present a taxonomy of solution concepts for multi-objective decision problems. Using this taxonomy, we discuss different solution methods. First, we assume that the model of the environment is known to the agents, leading to a *planning* setting. In Chapters 4 and 5, we discuss two different approaches for finding a *coverage set* using planning algorithms. In Chapter 6, we remove the assumption that the agents are given a model of the environment, and consider cases where they must learn about the environment through interaction. Finally, we discuss several illustrating applications in Chapter 6, followed by conclusions and future work in Chapter 8.

Diederik M. Roijers and Shimon Whiteson
April 2017

Acknowledgments

This book is based on our research on multi-objective decision making over the years. During this research, we collaborated with people whose input has been essential to our understanding of the field. We would like to thank several of them explicitly.

Together with Peter Vamplew and Richard Dazeley we wrote our 2013 survey article on multi-objective sequential decision making. The discussions we had about the nature of multi-objective decision problems were vital in shaping our ideas about this field, and lay the foundation for how we view multi-objective decision problems.

In the past few years, one of our main collaborators (and Diederik's other PhD supervisor), has been Frans A. Oliehoek. Together, we developed many algorithms for multi-objective decision making, including the CMOVE and OLS algorithms that we discuss in Chapters 4 and 5. Frans's vast expertise on partially observable decision problems and limitless capacity for generating new ideas have been invaluable to our work in the field of multi-objective decision making.

Together with Joris Scharpff, Matthijs Spaan, and Mathijs de Weerdt, we worked on the traffic network maintenance planning problem (which we discuss in Section 7.3), and in this context improved upon the original OLS algorithm (Chapter 5). We enjoyed this productive collaboration.

We would also like to thank our other past and present co-authors and collaborators who we have worked with on multi-objective decision making problems: Alexander Ihler, João Messias, Maarten van Someren, Chiel Kooijman, Maarten Inja, Maarten de Waard, Luisa Zintgraf, Timon Kanters, Philipp Beau, Richard Pronk, Carla Groenland, Elise van der Pol, Joost van Doorn, Daan Odijk, Maarten de Rijke, Ayumi Igarashi, Hossam Mossalam, and Yannis Assael.

Finally, we would like to thank several people with whom we had interesting discussions about multi-objective decision making over the years: Ann Nowé, Kristof van Moffaert, Tim Brys, Abdel-Illah Mouaddib, Paul Weng, Grégory Bonnet, Rina Dechter, Radu Marinescu, Shlomo Zilberstein, Kyle Wray, Patrice Perny, Paolo Viappiani, Pascal Poupart, Max Welling, Karl Tuyls, Francesco Delle Fave, Joris Mooij, Reyhan Aydoğan, and many others.

Diederik M. Roijers and Shimon Whiteson
April 2017

Table of Abbreviations

Abbreviation	Full Name	Location
AOLS	approximate optimistic linear support	Algorithm 5.10, Section 5.5
CCS	convex coverage set	Definition 3.7, Section 3.2.2
CH	convex hull	Definition 3.6, Section 3.2.2
CHVI	convex hull value iteration	Section 4.3.2
CLS	Cheng's linear support	Section 5.3
CMOVE	multi-objective variable elimination	Section 4.2.3
CoG	coordination graph	Definition 2.4, Section 2.2.1
CS	coverage set	Definition 3.5, Section 3.2
f	scalarization function	Definition 1.1, Section 1.1
MDP	Markov decision process	Definition 2.6, Section 2.3.1
MO-CoG	multi-objective coordination graph	Definition 2.5, Section 2.2.2
MODP	multi-objective decision problem	Definition 2.2, Section 2.1
MOMDP	multi-objective Markov decision process	Definition 2.8, Section 2.3.2
MORL	multi-objective reinforcement learning	Chapter 6
MOVE	multi-objective variable elimination	Algorithm 4.5, Section 4.2.3
MOVI	multi-objective value iteration	Section 4.3.2
OLS	optimistic linear support	Algorithm 5.8, Section 5.3
OLS-R	optimistic linear support with reuse	Algorithm 5.11, Section 5.6
PMOVI	Pareto multi-objective value iteration	Section 4.3.2
PCS	Pareto coverage set	Definition 3.11, Section 3.2.4
PMOVE	Pareto multi-objective variable elimination	Section 4.2.3
POMDP	partially observable Markov decision process	Section 5.2.1
PF	Pareto front	Definition 3.10, Section 3.2.4
SODP	single-objective decision problem	Definition 2.1, Section 2.1
U	undominated set	Definition 3.4, Section 3.2
VE	variable elimination	Algorithm 4.4, Section 4.2.1
VELS	variable elimination linear support	Section 5.7
VI	value iteration	Section 4.3.1
$\mathbf{V}^\pi$	value vector of a policy π	Definition 2.2, Section 2.1
Π	a set of allowed policies	Definition 2.1, Section 2.1
$\succ_P$	Pareto dominance relation	Definition 3.3, Section 3.1.2

CHAPTER 1

Introduction

Many real-world decision problems are so complex that they cannot be solved by hand. In such cases, *autonomous agents* that reason about these problems automatically can provide the necessary support for human decision makers. An agent is "anything that can be viewed as perceiving its environment through sensors and acting upon that environment through effectors" [Russell et al., 1995]. An artificial agent is typically a computer program—possibly embedded in specific hardware—that *takes actions* in an environment that changes as a result of these actions. Autonomous agents can act without human control or intervention, on a user's behalf [Franklin and Graesser, 1997].

Artificial autonomous agents can assist us in many ways. For example, agents can control manufacturing machines to produce products for a company [Monostori et al., 2006, Van Moergestel, 2014], drive a car in place of a human [Guizzo, 2011], trade goods or services on markets [Ketter et al., 2013, Pardoe, 2011], and help ensure security [Tambe, 2011]. As such, autonomous agents have enormous potential to improve our productivity and quality of life.

In order to successfully complete tasks, autonomous agents require the capacity to reason about their environment and the consequences of their actions, as well as the desirability of those consequences. The field of *decision theory* uses probabilistic models of the environment, called *decision problems*, to formalize the tasks about which such agents reason. Decision problems can include the *states* the environment can be in, the possible *actions* that agents can perform in each state, and how the state is affected by these actions. Furthermore, the desirability of actions and their effects are modeled as numerical feedback signals. These feedback signals are typically referred to as *reward*, *utility*, *payoff*, or *cost* functions. Solving a decision problem consists of finding a *policy*, i.e., rules for how to behave in each state, that is optimal in some sense with respect to these feedback signals.

In most research on planning and learning in decision problems, the desirability of actions and their effects are codified in a *scalar* reward function [Busoniu et al., 2008, Oliehoek, 2010, Thiébaux et al., 2006, Wiering and Van Otterlo, 2012]. In such scenarios, agents aim to maximize the expected (cumulative) reward over time.

However, many real-world decision problems have multiple objectives. For example, for a computer network we may want to maximize performance while minimizing power consumption [Tesauro et al., 2007]. Similarly, for traffic control, we may want to maximize throughput, minimize latency, maximize fairness to drivers, and minimize noise and pollution. In response to a query, we may want a search engine to provide a balanced list of documents that maximizes both the relevance to the query and the readability of the documents [Van Doorn et al., 2016]. In probabilistic planning, e.g., path planning for robots, we may want to maximize the probability of reaching a goal, while minimizing the expected cost of executing the plan [Bryce, 2008, Bryce et al., 2007]. Countless other real-world scenarios are naturally characterized by multiple objectives.

In all the cases mentioned above, the problem is more naturally expressed using a vector-valued reward function. When the reward function is vector-valued, the value of a policy is also vector-valued. Typically, there is no single policy that maximizes the value for all objectives simultaneously. For example, in a computer network, we can often achieve higher performance by using more power. If we do not know the exact preferences of the user with respect to these objectives, or indeed if these preferences may change over time, it can be crucial to produce a set of policies that offer different trade-offs between the objectives, rather than a single optimal policy.

The field of *multi-objective decision making* addresses how to formalize and solve decision problems with multiple objectives. This book provides an introductory overview of this field from the perspective of artificial intelligence. In addition to describing multi-objective decision problems and algorithms for solving them, we aim to make explicit the key assumptions that underly work in this area. Such assumptions are often left implicit in the multi-objective literature, which can be a source of confusion, especially for readers new to the topic. We also aim to synthesize these assumptions and offer a coherent, holistic view of the field.

We start by explicitly formulating the motivation for developing algorithms that are specific to multi-objective decision problems.

1.1   MOTIVATION

The existence of multiple objectives in a decision problem does not automatically imply that we require specialized multi-objective methods to solve it. On the contrary, if the decision problem can be *scalarized*, i.e., the vector-valued reward function can be converted to a scalar reward function, the problem may be solvable with existing single-objective methods. Such a conversion involves two steps [Roijers et al., 2013a]. The first step is to specify a *scalarization function* that expresses the utility of the user for different trade-offs between the objectives.

Definition 1.1   A *scalarization function* f is a function that maps a multi-objective value of a policy π of a decision problem, $\mathbf{V}^\pi$, to a scalar value $V_{\mathbf{w}}^\pi$:

$$V_{\mathbf{w}}^\pi = f(\mathbf{V}^\pi, \mathbf{w}),$$

where $\mathbf{w}$ is a weight vector that parameterizes f.

The second step of the conversion is to define a single-objective version of the decision problem such that the utility of each policy π equals the scalarized value of the original multi-objective decision problem $V_{\mathbf{w}}^\pi$.

Though it is rarely stated explicitly, all research on automated multi-objective decision making rests on the premise that there are decision problems for which performing one or both of these conversion steps is impossible, infeasible, or undesirable at the moment at which planning or learning must be performed. Here, we discuss three scenarios, illustrated in Figure 1.1, where this is the case, and specialized multi-objective methods are therefore preferable. In these scenarios, we assume a planning setting. However, in Chapter 6, we briefly address the learning setting.

Figure 1.1a depicts the *unknown weights scenario*. In this scenario, **w** is unknown at the moment when planning must occur: the *planning phase*. For example, consider a company that mines different resources from different mines spread out through a mountain range. The people who work for the company live in villages at the foot of the mountains. In Figure 1.2, we depict the problem this company faces: in the morning, one van per village needs to transport workers from that village to a nearby mine, where various resources can be mined. Different mines yield different quantities of resource per worker. The market prices per unit of resource vary through a stochastic process and every price change can affect the optimal assignment of vans. Furthermore, the expected price variation increases with time. It is therefore critical to act based on the latest possible price information in order to maximize performance.

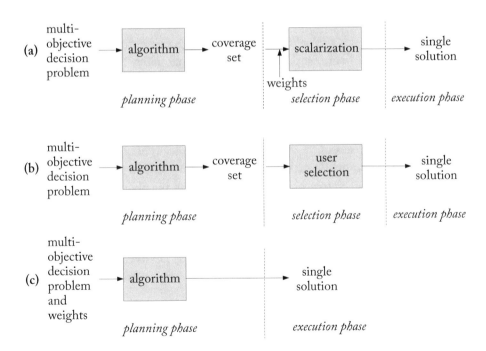

Figure 1.1: The three motivating scenarios for multi-objective decision-theoretic planning: (a) the unknown weights scenario, (b) the decision support scenario, and (c) the known weights scenario.

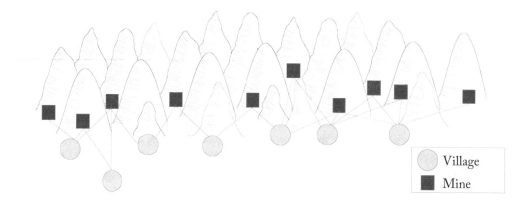

Figure 1.2: Mining company example.

Because computing the optimal van assignment takes time, computing the optimal assignment with each price change is impossible. Therefore, we need a multi-objective method that computes a set containing an optimal solution for every possible value of the prices, $\mathbf{w}$. We call such a set a *coverage set*, as it "covers" all possible preferences of the user (i.e., the possible prices) with respect to the objectives (as specified by f).[1] Although computing a coverage set is computationally more expensive than computing a single optimal policy for a given price, it needs to be done only once. Furthermore, the *planning phase* can take place in advance, when more computational resources are available.

In the *selection phase*, when the prices ($\mathbf{w}$) are revealed and we want to use as little computation time as possible, we can use the coverage set to determine the best policy by simple maximization. Finally, in the *execution phase*, the selected policy is employed.

In this book, we focus on algorithms for the planning/learning phase. For the selection phase, specialized *preference elicitation* algorithms for selecting the policy with the optimal utility from the coverage set may be necessary when the coverage set is large. For example, Chu and Ghahramani [2005], propose an approach to preference elicitation via Gaussian processes.

In the unknown weights scenario, *a priori* scalarization is impossible, because it would shift the burden of computation toward a point in time where it is not available. The scalarization f is known, and the weights $\mathbf{w}$ will become available in the selection phase, where a single policy is selected for execution.

By contrast, in the *decision support scenario* (Figure 1.1b), $\mathbf{w}$ and f are never made explicit. In fact, scalarization is infeasible throughout the entire decision-making process because of the difficulty of specifying $\mathbf{w}$ and/or f. For example, when a community is considering the construction of a new metro line, economists may not be able to accurately compute the economic benefit of reduced commuting times. The users may also have "fuzzy" preferences that defy meaningful

---

[1]We provide a formal definition of the term *coverage set* in Chapter 3, Definition 3.5.

quantification. For example, if construction of the new metro line could be made more efficient by building it in such a way that it obstructs a beautiful view, then a human designer may not be able to quantify the loss of beauty. The difficulty of specifying the exact scalarization is especially apparent when the designer is not a single person but a committee or legislative body whose members have different preferences and agendas, such as the politicians and interest groups involved in constructing the metro line. In such a system, the multi-objective planning method is used to calculate a coverage set with respect to the constraints that can safely be imposed on f and $\mathbf{w}$. For example, we can safely assume that gaining value in one objective, without reducing the value in any of the others cannot reduce the utility to the user (i.e., the scalarized value).

As shown in Figure 1.1b, the decision support scenario proceeds similarly to the unknown weights scenario in the planning phase. In the selection phase, however, the user or users directly select a policy from the coverage set according to their idiosyncratic preferences, rather than explicitly computing a numerical utility by applying the scalarization function to each value vector.

In the decision support scenario, one could still argue that scalarization before planning (or learning) is possible in principle. For example, the loss of beauty can be quantified by measuring the resulting drop in housing prices in neighborhoods that previously enjoyed an unobstructed view. However, the difficulty with explicit scalarization is not only that doing so may be impractical but, more importantly, that it forces the users to express their preferences in a way that may be inconvenient and unnatural. This is because selecting $\mathbf{w}$ requires weighing hypothetical trade-offs, which can be much harder than choosing from a set of actual alternatives. This is a well understood phenomenon in the field of *decision analysis* [Clemen, 1997], where the standard workflow involves presenting alternatives *before* soliciting preferences. In the same way, algorithms for multi-objective decision problems can provide critical decision support; rather than forcing the users to specify f and $\mathbf{w}$ in advance, these algorithms just prune policies that would not be optimal for any f and $\mathbf{w}$ that fit the known constraints on the preferences of the users, and produce a coverage set. Because this coverage set contains optimal solutions for all f and $\mathbf{w}$ that fit the known constraints—rather than just all $\mathbf{w}$ for a specified f, as in the unknown weights scenario—it offers a range of alternatives from which the users can select, according to preferences whose relative importance is not easily quantified.

Finally, we present one more scenario, which we call the *known weights scenario*, that requires explicit multi-objective methods. In this scenario, we assume that $\mathbf{w}$ is known at the time of planning and thus scalarization is possible. However, it may be *undesirable* because of the difficulty of the second step in the conversion. In particular, if f is nonlinear, then the resulting single-objective problem may be much more complex than the original multi-objective problem. As a result, finding the optimal policy may be intractable while the original multi-objective problem is tractable. For example, in multi-objective Markov decision processes (MOMDPs[2]), which

---

[2]This abbreviation is also used for *mixed-observability MDPs* [Ong et al., 2010], which we do not consider here; we use the abbreviation MOMDPs solely for multi-objective MDPs.

we formally define in Chapter 2, nonlinear scalarization destroys the *additivity property* on which single-objective solution methods rely (see Section 3.2.3).

Figure 1.1c depicts the known weights scenario. In contrast to the other scenarios, in the known weights scenario, the multi-objective method only produces one policy, which is then executed, i.e., there is no separate selection phase.

The scenarios we have presented here require explicit multi-objective methods because *a priori* scalarization of the multi-objective decision problems, and subsequent solving with standard single-objective methods, does not apply. In this book, we focus on the two multi-policy scenarios, i.e., the unknown weights and decision support scenarios, in which the goal of a multi-objective planning method is to produce a coverage set. From this coverage set, the policy that maximizes *user utility* is selected in the selection phase.

Of course, computing a coverage is in general more difficult than finding a single policy, and thus multi-objective methods are typically more expensive than their single-objective counterparts. It is natural then to wonder whether the complications of a multi-objective approach are merited. After all, many single-objective problems are already intractable. In this book, we take a pragmatic perspective: multi-objective methods are not a panacea and are not always the best option, even if the problem is naturally modeled with multiple objectives. If the scalarization weights are known (or can be reasonably estimated) before planning begins, and *a priori* scalarization does not yield an intractable problem, then converting a multi-objective problem to a single-objective one may be the best option. However, in any of the many cases where such a conversion is not possible or practical, then the multi-objective methods discussed in this book may prove indispensable.

1.2   UTILITY-BASED APPROACH

The goal of solving all—including multi-objective—decision problems is to maximize user utility. However, in the unknown weights and decision support scenarios, we cannot optimize user utility directly because, at the time when planning or learning takes place, the parameters $\mathbf{w}$ of the scalarization function f, which maps the multi-objective values to a scalar utility, are unknown. Therefore, we must compute a *coverage set*: a set of policies such that, for every possible scalarization, a maximizing policy is in the set (see Definition 3.5).

In this book, we argue that which policies should be included in the coverage set should be derived from what we know about f. We call this the *utility-based approach*, in contrast to the *axiomatic approach* that axiomatically assumes the coverage set is the *Pareto front*, which we define formally in Chapter 3. In short, the Pareto front is the set of all policies for which there is no other policy that has at least equal value in all objectives and has a higher value in at least one objective. Indeed, the Pareto front contains at least one optimal policy for most, if not all, scalarization functions that occur in practice. However, we argue that, while the Pareto front is *sufficient*, it is often not *necessary*. In fact, we show in Chapter 3 that the full Pareto front is required only in the special case where the scalarization function is nonlinear and policies must be deterministic.

A utility-based approach thus often results in a much smaller coverage set, which is typically less expensive to compute and easier for the user to examine.

Another advantage of the utility-based approach is that it is possible to derive how much utility is maximally lost when an exact coverage set cannot be computed [Zintgraf et al., 2015]. Such bounds are often crucial for a meaningful interpretation of the quality of approximate methods for decision-theoretic planning or learning, especially when comparing algorithms [Oliehoek et al., 2015]. Furthermore, the bounds provide insight into the quality and reliability of the selected final policy.

CHAPTER 2

Multi-Objective Decision Problems

In this chapter, we introduce the concept of a multi-objective decision problem. Then we describe two concrete classes of multi-objective decision problems that we use throughout the book: multi-objective coordination graphs and multi-objective Markov decision processes. However, before introducing the concrete multi-objective decision problems, we first introduce their single-objective counterparts.

2.1 MULTIPLE OBJECTIVES

In this book, we focus on different (cooperative) multi-objective decision problems. Multi-objective decision problems contrast single-objective decision problems, which are more common in the literature. In their most general form, single-objective decision problems can be defined as a set of policies and a value function that we wish to maximize:

Definition 2.1 A cooperative *single-objective decision problem (SODP)*, consists of:

- a set of allowed (joint) *policies* Π,

- a value function that assigns a real numbered value, $V^\pi \in \mathbb{R}$, to each joint policy $\pi \in \Pi$, corresponding to the desirability, i.e., the utility, of the policy.

Definition 2.2 In a cooperative *multi-objective decision problem (MODP)*, Π is the same as in an SODP, but

- there are $d \geq 1$ *objectives*, and

- the value function assigns a *value vector*, $\mathbf{V}^\pi \in \mathbb{R}^d$, to each joint policy $\pi \in \Pi$, corresponding to the desirability of the policy *with respect to each objective*.

We denote the value of policy π in the i-th objective as V_i^π.

Both $\mathbf{V}^\pi$ and Π may have particular forms that affect the way they should be computed, as we discuss in Chapter 3. For example, there may be multiple agents, each with its own action space. In such settings, a solution consists of a *joint policy* containing a *local policy* for each agent.

We assume that Π is known and that it is in principle possible to compute the value of each (joint) policy. Furthermore, we assume that, if there are multiple agents, these agents are *cooperative*.

Definition 2.3   A multi-agent MODP is *cooperative* if and only if all agents get the same (team) value, $\mathbf{V}^{\pi}$, for executing a joint policy $\pi \in \Pi$, i.e., there are no individual rewards. A single-agent MODP is cooperative by default.

This definition of cooperative is common in the field of decision theory, e.g., in multi-agent MDPs [Boutilier, 1996, Scharpff et al., 2016] and Dec-POMDPs [Oliehoek and Amato, 2016]. However, the term "cooperative" is used differently in cooperative game theory [Chalkiadakis et al., 2011, Igarashi and Roijers, 2017], in which agents form coalitions on the basis of their individual utilities. In this book, we consider only decision problems that are cooperative according to Definition 2.3.

In an SODP, the value function provides a complete ordering on the joint policies, i.e., for each pair of policies π and π', V^{π} must be greater than, equal to, or less than $V^{\pi'}$. By contrast, in an MODP, the presence of multiple objectives means that the value function $\mathbf{V}^{\pi}$ is a vector rather than a scalar. Such value functions supply only a partial ordering. For example, it is possible that, $V_i^{\pi} > V_i^{\pi'}$ but $V_j^{\pi} < V_j^{\pi'}$. Consequently, unlike in an SODP, we can no longer determine which values are optimal without additional information about how to prioritize the objectives, i.e., about what the *utility* of the user is for different trade-offs between the objectives.

In the *unknown weights* and *decision support scenarios* (Figure 1.1), the parameters of the scalarization function $\mathbf{w}$, or even f itself, are unknown during the planning or learning phases. Therefore, an algorithm for solving an MODP should return a set of policies that contains an optimal policy for each possible $\mathbf{w}$. Given such a solution set, the user can pick the policy that maximizes her utility in the *selection phase*. We want the solution set to contain at least one optimal policy for every possible scalarization (in order to guarantee optimality), but we also want the solution set to be as small as possible, in order to make the selection phase as efficient as possible. We discuss which solution sets are optimal, and how this can be derived from different assumptions about the scalarization function f (Definition 1.1), and the set of permitted policies Π in the MODP in Chapter 3. In the rest of this section, we introduce two different MODP problem classes.

2.2   MULTI-OBJECTIVE COORDINATION

The first class of MODPs that we treat is the *multi-objective coordination graph (MO-CoG)*.[1] In a MO-CoG, multiple agents need to coordinate their actions in order to be effective. For example, in the mining problem of Figure 1.2, each agent represents a van with workers from a single village. Each of these vans can go to different mines within reasonable traveling distance, leading

---

[1]In the literature, MO-CoGs have many different names: *multi-objective weighted constraint satisfaction problems (MO-WCSPs)* [Rollón, 2008], *multi-objective constraint optimization problems (MO-COPs)* [Marinescu, 2011], and *multi-objective collaborative graphical games (MO-CoGG)* [Roijers et al., 2013c].

to a set of different possible actions for each agent. Each mine yields a different expected amount of gold (the first objective) and silver (the second objective). Because mining can be done more efficiently when more workers are present at a mine, it is vitally important that the different agents (i.e., vans) coordinate which mines they go to.

Other examples of problems that can be modeled as a MO-CoG are: risk-sensitive combinatorial auctions, in which we want to maximize the total revenue, while minimizing the risk for the auctioneer [Marinescu, 2011], and maintenance scheduling for offices in which the energy consumption, costs, and overtime for the maintenance staff must all be minimized [Marinescu, 2011].

2.2.1 SINGLE-OBJECTIVE COORDINATION GRAPHS

Before we formally define MO-CoGs, we first define the corresponding single-objective problem, i.e., *coordination graphs* (CoGs) [Guestrin et al., 2002, Kok and Vlassis, 2004]. In the context of coordination graphs, the notion of reward is typically referred to as *payoff* in the literature. Payoff is usually denoted u (for *utility*). We adopt this terminology and notation.

Definition 2.4 A *coordination graph* (CoG) [Guestrin et al., 2002, Kok and Vlassis, 2004] is a tuple $\langle \mathcal{D}, \mathcal{A}, \mathcal{U} \rangle$, where

- $\mathcal{D} = \{1, \ldots, n\}$ is the set of n agents,

- $\mathcal{A} = \mathcal{A}_i \times \ldots \times \mathcal{A}_n$ is the joint action space: the Cartesian product of the finite action spaces of all agents. A joint action is thus a tuple containing an action for each agent $\mathbf{a} = \langle a_1, \ldots, a_n \rangle$, and

- $\mathcal{U} = \{u^1, \ldots, u^\rho\}$ is the set of ρ scalar *local payoff functions*, each of which has limited *scope*, i.e., it depends on only a subset of the agents. The total team payoff is the sum of the local payoffs: $u(\mathbf{a}) = \sum_{e=1}^{\rho} u^e(\mathbf{a}_e)$.

In order to ensure that the coordination graph is *fully cooperative*, all agents share the payoff function $u(\mathbf{a})$. We abuse the notation e both to index a local payoff function u^e and to denote the subset of agents in its scope; $\mathbf{a}_e$ is thus a *local joint action*, i.e., a joint action of this subset of agents. The decomposition of $u(\mathbf{a})$ into local payoff functions can be represented as a *factor graph* [Bishop, 2006] (Figure 2.1); a bipartite graph containing two types of vertices: agents (variables) and local payoff functions (factors), with edges connecting local payoff functions to the agents in their scope.

The main challenge in a CoG is that the size of the joint action space $\mathcal{A}$ grows exponentially with the number of agents. It thus quickly becomes intractable to enumerate all joint actions and their associated payoffs. Key to solving CoGs is therefore to exploit *loose couplings* between agents, i.e., each agent's behavior directly affects only a subset of the other agents.

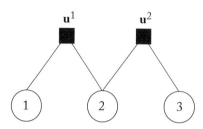

Figure 2.1:   A CoG with three agents and two local payoff functions. The factor graph illustrates the loose couplings that result from the decomposition into local payoff functions. In particular, each agent's choice of action directly depends only on those of its immediate neighbors, e.g., once agent 1 knows agent 2's action, it can choose its own action without considering agent 3.

Figure 2.1 shows the factor graph of an example CoG in which the team payoff function decomposes into two local payoff functions, each with two agents in scope:

$$u(\mathbf{a}) = \sum_{e=1}^{\rho} u^e(\mathbf{a}_e) = u^1(a_1, a_2) + u^2(a_2, a_3).$$

The local payoff functions are defined in Table 2.1. We use this CoG as a running example throughout this book. The local payoff functions, with their limited scopes, encode the loose couplings: each agent can only directly affect another agent when they share, i.e., are both in the scope of, a local payoff function. For example, if we fix the action for agent 2 to be $\dot{a}_2$, then agents 1 and 3 can decide upon their optimal actions independently, as they do not directly affect each other.

Table 2.1: The payoff matrices for $u^1(a_1, a_2)$ (left) and $u^2(a_2, a_3)$ (right). There are two possible actions per agent, denoted by a dot ($\dot{a}_1$) and a bar ($\bar{a}_1$).

	$\dot{a}_2$	$\bar{a}_2$
$\dot{a}_1$	3.25	0
$\bar{a}_1$	1.25	3.75

	$\dot{a}_3$	$\bar{a}_3$
$\dot{a}_2$	2.5	1.5
$\bar{a}_2$	0	1

2.2.2   MULTI-OBJECTIVE COORDINATION GRAPHS

We now consider the multi-objective setting:

Definition 2.5   A *multi-objective coordination graph (MO-CoG)* [Roijers et al., 2015b] is a tuple $\langle \mathcal{D}, \mathcal{A}, \mathcal{U} \rangle$ where:

- $\mathcal{D}$ and $\mathcal{A}$ are the same as in a CoG, but,

- $\mathcal{U} = \{\mathbf{u}^1, ..., \mathbf{u}^\rho\}$ is now the set of ρ, d-*dimensional* local payoff functions. The total team payoff is the sum of local *vector-valued* payoffs: $\mathbf{u}(\mathbf{a}) = \sum_{e=1}^{\rho} \mathbf{u}^e(\mathbf{a}_e)$.

For example, payoffs for a MO-CoG structured as in Figure 2.1, i.e.,

$$\mathbf{u}(\mathbf{a}) = \mathbf{u}^1(a_1, a_2) + \mathbf{u}^2(a_2, a_3),$$

are provided in Table 2.2.

Table 2.2: The two-dimensional payoff matrices for $\mathbf{u}^1(a_1, a_2)$ (left) and $\mathbf{u}^2(a_2, a_3)$ (right)

	$\dot{a}_2$	$\bar{a}_2$
$\dot{a}_1$	(4, 1)	(0,0)
$\bar{a}_1$	(1, 2)	(3, 6)

	$\dot{a}_3$	$\bar{a}_3$
$\dot{a}_2$	(3, 1)	(1, 3)
$\bar{a}_2$	(0, 0)	(1, 1)

The only difference between a MO-CoG and a CoG is the dimensionality of the payoff functions. A CoG is thus a special case of a MO-CoG, i.e., a MO-CoG in which $d = 1$. Furthermore, when preferences are known, it may be possible to scalarize a MO-CoG and thus transform it into a CoG. For example, if we know the scalarization function is linear, i.e., $f(\mathbf{u}, \mathbf{w}) = \mathbf{w} \cdot \mathbf{u}$, and its parameter vector $\mathbf{w} = (0.75, 0.25)$ is, then we can scalarize the multi-objective payoff functions of Table 2.2 to the single-objective payoff functions of Table 2.1 before planning.

In a *deterministic policy* for a MO-CoG, the agents pick one joint action. In a *stochastic policy*, the agents draw a joint action from a probability distribution. Note that such a probability distribution is in general defined over *joint* actions, and there can thus still be coordination between the agents when the policy is stochastic.

When f and/or $\mathbf{w}$ are unknown in the planning or learning phase—as is the case in the *unknown weights* and *decision support* scenarios discussed in Section 1.1—we need to produce a set of policies that contains at least one optimal policy for each possible f and $\mathbf{w}$. The solution of a MO-CoG is thus a coverage set of policies. In general, this can contain both deterministic and stochastic policies. We explain why this is important for MO-CoGs (but not for CoGs) in Chapter 3.

2.3   MULTI-OBJECTIVE MARKOV DECISION PROCESSES

The second class of MODPs that we treat is the *multi-objective Markov decision process (MOMDP)*, in which a single agent needs to perform a sequence of actions over time. This sequence of actions typically takes place in an environment that is affected by these actions. Therefore, the agent has to consider not only its immediate reward, but also the reward it will be able obtain later in the states it visits in the future.

Consider the robot (shown as a Pacman symbol) in Figure 2.2 that needs to navigate in a socially appropriate way to reach the blue target zone in the upper right corner. We want the robot to reach the target as soon as possible, i.e., minimize the time to reach the target, but also minimize the hindrance that the robot causes to the other person by avoiding her personal space (indicated in red) along the way. By driving through the personal space of the person, it can obtain a higher value with respect to the first objective but a lower value in the second objective. Which policy is optimal thus depends on how much we value the first objective relative to the second objective.

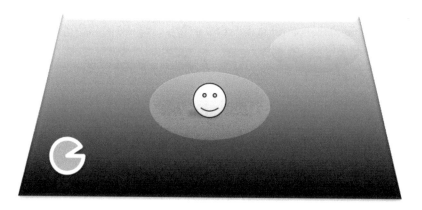

Figure 2.2: A social robot navigation problem MDP: a robot (indicated by the pacman symbol) needs to reach the target zone in the upper right corner (objective 1). However, the robots needs to avoid the personal space of the person standing in between the start position of the robot and the target zone (objective 2).

2.3.1   SINGLE-OBJECTIVE MARKOV DECISION PROCESSES

The single-objective version of an MOMDP is an MDP:

Definition 2.6   A (single-objective) *Markov decision process (MDP)* [Bellman, 1957b] is a tuple $\langle S, A, T, R, \mu_0 \rangle$ where,

- S is the state space, i.e., the set of possible states the environment can be in,

- A is the action space, i.e., the set of actions the agent can take,

- $T : S \times A \times S \rightarrow [0, 1]$ is the transition function, giving the probability of a next state given an action and a current state,

- $R : S \times A \times S \rightarrow \mathbb{R}$ is the reward function, specifying the immediate expected scalar reward corresponding to a transition.

- μ_0 is the distribution over initial states s_0.

At each timestep t, the agent observes the current state of the environment $s \in \mathcal{S}$. When the agent takes an action $a \in \mathcal{A}$ the environment transitions to a new state s'.

The state in an MDP is *Markovian*, i.e., the current state s of the environment and the current action of the agent a are a sufficient statistic for predicting the next transition probabilities $T(s'|s, a)$ and the associated expected immediate reward. The agent's history, i.e., the states and actions that led to the current state, do not provide additional information in that respect.

The agent's goal in an MDP is to find a policy π that maximizes the return R_t, which is some function of the rewards received from timestep t and onward. In the broadest sense, a policy can condition on everything that is known to the agent.

A *state-independent value function* V^π specifies the expected return when following π from the initial state:

$$V^\pi = E[R_0 \mid \pi, \mu_0]. \tag{2.1}$$

We further specify first the return, R_t, and then the policy, π.

Typically, the return is *additive* [Boutilier et al., 1999], i.e., it is a sum of the rewards attained at each timestep. When the returns are additive, V^π becomes an expectation over this sum. In a *finite-horizon setting* there is a limited number of timesteps, h, and the sum is typically undiscounted:

$$V^\pi = E[\sum_{t=0}^{h-1} R(s_t, a_t, s_{t+1})|\pi, \mu_0]. \tag{2.2}$$

In a *discounted infinite-horizon setting*, the number of timesteps is not limited, but there is a discount factor, $0 \leq \gamma < 1$, that specifies the relative importance of future rewards with respect to immediate rewards:

$$V^\pi = E[\sum_{t=0}^{\infty} \gamma^t R(s_t, a_t, s_{t+1})|\pi, \mu_0]. \tag{2.3}$$

Which action a_t is chosen by the agent at each timestep t depends on its policy π. If the policy is *stationary*, i.e., it conditions only on the current state, then it can be formalized as $\pi : S \times A \rightarrow [0, 1]$: it specifies, for each state and action, the probability of taking that action in that state. We can then specify the *state value function* of a policy π:

$$V^\pi(s) = E[R_t \mid \pi, s_t = s],$$

for all t when $s_t = s$. The *Bellman equation* restates this expectation recursively for stationary policies:

$$V^\pi(s) = \sum_a \pi(s, a) \sum_{s'} T(s, a, s')[R(s, a, s') + \gamma V^\pi(s')]. \tag{2.4}$$

Note that the Bellman equation, which forms the heart of most standard solution algorithms such as *dynamic programming* [Bellman, 1957a] and *temporal difference methods* [Sutton and Barto,

1998], explicitly relies on the assumption of additive returns. This is important because nonlinear scalarization functions f can interfere with this additivity property, making planning and learning methods that rely on the Bellman equation not directly applicable, as we discuss in Section 3.2.3.

State value functions induce a partial ordering over policies, i.e., π is better than or equal to π' if and only if its value is greater for all states:

$$\pi \succeq \pi' \Leftrightarrow \forall s, V^\pi(s) \geq V^{\pi'}(s).$$

A special case of a stationary policy is a *deterministic* stationary policy, in which one action is chosen with probability 1 for every state. A deterministic stationary policy can be seen as a mapping from states to actions: $\pi : S \to A$. For single-objective MDPs, there is always at least one *optimal* policy π, i.e., $\forall \pi' : \pi \succeq \pi'$, that is stationary and deterministic.

Theorem 2.7   *For any additive infinite-horizon single-objective MDP, there exists a deterministic stationary optimal policy [Boutilier et al., 1999, Howard, 1960].*

If more than one optimal policy exists, they share the same value function, known as the *optimal value function* $V^*(s) = \max_\pi V^\pi(s)$. The *Bellman optimality equation* defines the optimal value function recursively:

$$V^*(s) = \max_a \sum_{s'} T(s,a,s')[R(s,a,s') + \gamma V^*(s')]. \tag{2.5}$$

Note that, because it maximizes over actions, this equation makes use of the fact that there is an optimal deterministic stationary policy. Because an optimal policy maximizes the value for every state, such a policy is optimal regardless of the initial state distribution μ_0. However, the state-independent value (Equation 2.1) can be different for different initial state distributions. Using μ_0, the state value function can be translated back into the state-independent value function (Equation 2.1):

$$V^\pi = \sum_{s \in S} \mu_0(s) V^\pi(s).$$

2.3.2   MULTI-OBJECTIVE MARKOV DECISION PROCESSES

In many decision problems, such as the social robot in Figure 2.2, it is impossible, undesirable, or infeasible to define a scalar reward function, and we need a vector-valued reward function, leading to an MOMDP.

Definition 2.8   A *multi-objective Markov decision process (MOMDP)* [Roijers et al., 2013a] is a tuple $\langle S, A, T, R \rangle$ where,

- S, A, and T are the same as in an MDP, but,

- $\mathbf{R} : S \times A \times S \to \mathbb{R}^d$ is now a d-dimensional reward function, specifying the expected immediate *vector-valued* reward corresponding to a transition.

MOMDPs have recently been applied to many real-world decision problems, including: water reservoir control [Castelletti et al., 2013, 2008, Giuliani et al., 2015], where policies for releasing water from a dam must be found while balancing multiple uses of the reservoir, including hydroelectric production and flood mitigation; office building environmental control [Kwak et al., 2012], in which energy consumption must be minimized while maximizing the comfort of the building's occupants; and medical treatment planning [Lizotte et al., 2010, 2012], in which the effectiveness of the treatment must be maximized, while minimizing the severity of the side effects.

In an MOMDP, when an agent executes a policy π, its value, $\mathbf{V}^{\pi}$ is vector-valued, as it is an expectation of the sum over vector-valued rewards, i.e.,

$$\mathbf{V}^{\pi} = E[\sum_{t=0}^{h} \mathbf{R}(s_t, a_t, s_{t+1})|\pi, \mu_0],$$

in the *finite-horizon setting*, and,

$$\mathbf{V}^{\pi} = E[\sum_{t=0}^{\infty} \gamma^t \mathbf{R}(s_t, a_t, s_{t+1})|\pi, \mu_0].$$

in the *infinite horizon setting*.

In an MOMDP, we must find a coverage set of policies. This is significantly harder than in an MDP, not only because we need to produce a coverage set, but also because several assumptions that hold in a single-objective MDP do not hold for MOMDPs in general, requiring us to consider a larger search space of policies. In the next chapter, we provide an overview of which types of policies we need to consider given different assumptions about the MOMDP.

CHAPTER 3

Taxonomy

So far, we have motivated multi-objective decision making and described two specific multi-objective decision problems. In this chapter, we discuss what constitutes an optimal solution. Unfortunately, there is no simple answer to this question, as it depends on several critical factors.

Recall that, in Chapter 1, we proposed a *utility-based* approach for deciding what constitutes an optimal solution to a multi-objective decision problem. Rather than axiomatically assuming that the Pareto front is the appropriate solution concept, we advocate for deriving the solution concept from what we know about the problem and the scalarization function.

In this chapter, we identify the appropriate solution concept for a number of multi-objective settings. In particular, we propose a taxonomy that organizes multi-objective problems in terms of their underlying assumptions. For the sake of brevity, the taxonomy focuses on the MOMDP introduced in Chapter 2. Near the end of this chapter, however, we briefly discuss how the same principles apply to MO-CoGs.

The goal of our taxonomy is to cover most research on multi-objective decision making while remaining simple and intuitive. However, some research does not fit neatly in our taxonomy. We discuss such settings in Section 3.5.

3.1 CRITICAL FACTORS

We start by identifying three critical factors that affect what constitutes an optimal solution to a multi-objective problem. After identifying these factors, we show in Section 3.2 how they lead to different solution concepts.

3.1.1 SINGLE VS. MULTIPLE POLICIES

Following the approach of Vamplew et al. [2011], we first distinguish problems in which only one policy is sought from ones in which multiple policies are sought. Which case holds depends on which of the three motivating scenarios discussed in Chapter 1 we face.

In the unknown weights and decision support scenarios, the solution consists of multiple policies. Though these two scenarios are conceptually quite different, from an algorithmic perspective they are identical. The reason is that they are both characterized by a strict separation of the decision-making process into two phases: the planning or learning phase and the execution phase (although in on-line settings, the agent may go back and forth between the two).

In the known weights scenario, **w** is known before planning or learning begins. Therefore, returning multiple policies is unnecessary. However, as mentioned in Chapter 1, scalarization can yield a single-objective MDP that is difficult to solve.

3.1.2 LINEAR VS. MONOTONICALLY INCREASING SCALARIZATION FUNCTIONS

The second critical factor is the nature of the scalarization function. In this section, we discuss two types of scalarization function: those that are linear combinations of the rewards and those that are merely monotonically increasing functions of them.

A common assumption about the scalarization function, e.g., Barrett and Narayanan [2008], Natarajan and Tadepalli [2005], is that f is linear, i.e., it computes the weighted sum of the values for each objective.

Definition 3.1 A linear scalarization function computes the inner product of a weight vector **w** and a value vector $\mathbf{V}^\pi$:

$$V_{\mathbf{w}}^\pi = \mathbf{w} \cdot \mathbf{V}^\pi. \tag{3.1}$$

Each element of **w** specifies how much one unit of value for the corresponding objective contributes to the scalarized value. For convenience, we typically assume that f computes not only a linear combination but a convex one, i.e., the elements of the weight vector **w** are all positive real numbers and constrained to sum to 1.

Linear scalarization functions are a simple and intuitive way to scalarize. One common situation in which they are applicable is when rewards can be easily translated into monetary value. For example, consider the mining task mentioned in Chapter 1, in which different policies yield different expected quantities of various minerals. If the prices per kilo of those minerals fluctuate daily, then the task can be formulated as a multi-objective problem, with each objective corresponding to a different mineral. Each element of $\mathbf{V}^\pi$ then reflects the expected number of kilos of that mineral that are mined under π and the scalarized value $V_{\mathbf{w}}^\pi$ corresponds to the monetary value of everything that is mined. $V_{\mathbf{w}}^\pi$ can be computed only when **w**, corresponding to the (normalized) current price per kilo of each mineral, becomes known.

While linear scalarization functions are intuitive and simple, they are not always adequate for expressing the user's preferences. For example, suppose there are two minerals that can be mined and only three policies are available: π_1 sends the mining equipment to a location where only the first mineral can be mined, π_2 to a location where only the second mineral can be mined, and π_3 to a location where both minerals can be mined. Suppose the owner of the equipment prefers π_3, e.g., because it at least partially appeases clients with different interests. However, it may be the case that, because the location corresponding to π_3 has fewer minerals, there is no linear scalarization function for which π_3 is optimal. Thus, the owner's preference of π_3 implies that she, implicitly or explicitly, employs a nonlinear scalarization function.

Here, we consider the case in which f can be nonlinear, and corresponds to a common notion of the relationship between reward and utility. This class of possibly nonlinear scalarizations are the *strictly monotonically increasing* scalarization functions. These functions adhere to the constraint that, if a policy is changed in such a way that its value increases in one or more of the objectives, without decreasing in any other objectives, then the scalarized value also increases.

Definition 3.2   A scalarization function f is *strictly monotonically increasing* if:

$$(\forall i, \ V_i^\pi \geq V_i^{\pi'} \wedge \exists i, \ V_i^\pi > V_i^{\pi'}) \Rightarrow (\forall \mathbf{w}, \ V_\mathbf{w}^\pi > V_\mathbf{w}^{\pi'}). \tag{3.2}$$

For the sake of brevity, in the remainder of this book we refer to strictly monotonically increasing scalarization functions simply as monotonically increasing. Linear scalarization functions (with strictly positive weights) are included in this class of functions. The condition on the left-hand side of (3.2) is more commonly known as *Pareto dominance* [Pareto, 1896].

Definition 3.3   A policy π *Pareto-dominates* another policy π' when its value is at least as high in all objectives and strictly higher in at least one objective:

$$\mathbf{V}^\pi \succ_P \mathbf{V}^{\pi'} \Leftrightarrow \forall i, V_i^\pi \geq V_i^{\pi'} \wedge \exists i, V_i^\pi > V_i^{\pi'}. \tag{3.3}$$

Demanding that f is monotonically increasing is quite a minimal constraint, as it requires only that, all other things being equal, getting more reward for a certain objective is always better. In fact, it is difficult to think of any f that violates this constraint without employing a highly unnatural notion of reward.

3.1.3   DETERMINISTIC VS. STOCHASTIC POLICIES

The third critical factor is whether only deterministic policies are considered or stochastic ones are also allowed. While in most applications there is no reason to exclude stochastic policies *a priori*, there can be cases when stochastic policies are clearly undesirable or even unethical. For example, if the policy determines the clinical treatment of a patient [Lizotte et al., 2010, Shortreed et al., 2011], then flipping a coin to determine the course of action may be inappropriate. We denote the set of deterministic policies Π_D and the set of stationary policies Π_S. Both sets are subsets of all policies: $\Pi_D \subseteq \Pi \wedge \Pi_S \subseteq \Pi$. Finally, the set of policies that are both deterministic and stationary is the intersection of both these sets, denoted $\Pi_{DS} = \Pi_D \cap \Pi_S$.

In single-objective problems, these factors are typically not critical. For example, in MDPs, thanks to Theorem 2.7, we can restrict our search to deterministic stationary policies, i.e., the optimal attainable value is attainable with a deterministic stationary policy: $\max_{\pi \in \Pi} V^\pi = \max_{\pi \in \Pi_{DS}} V^\pi$. However, the situation is more complex in multi-objective settings. As we show below, in some cases, forbidding stochastic policies can affect the nature of the appropriate solution concept.

3.2   SOLUTION CONCEPTS

The three critical factors defined above can be used to organize multi-objective settings into a taxonomy, as shown in Table 3.1. Each cell indicates the solution concept that is appropriate for the corresponding setting. In this section, we step through all the cells in the taxonomy and derive the given solution concept. Before doing so, however, we start with two definitions that, in a generic way, formalize the notion of a solution to any multi-objective problem for which multiple policies are required.

Table 3.1: The multi-objective problem taxonomy showing the critical factors and the nature of the resulting optimal solution. The columns describe whether the problem necessitates a single policy or multiple ones, and whether those policies must be deterministic (by specification) or are allowed to be stochastic. The rows describe whether the scalarization function is a linear combination of the rewards or whether this cannot be assumed and the scalarization function is merely a monotonically increasing function of them. The contents of each cell describe what an optimal solution for the given setting looks like.

	Single Policy (known weights)		Multiple Policies (unknown weights or decision support)	
	Deterministic	Stochastic	Deterministic	Stochastic
Linear Scalarization	One deterministic stationary policy (1)		Convex coverage set of deterministic stationary policies (2)	
Monotonically Increasing Scalarization	One deterministic non-stationary policy (3)	One mixture policy of two or more deterministic stationary policies (4)	Pareto coverage set of deterministic non-stationary policies (5)	Convex coverage set of deterministic stationary policies (6)

Recall that, in the unknown weights and decision support scenarios, **w** is unavailable in the planning or learning phase. Consequently, a multi-objective algorithm for such settings should in general return a set containing multiple policies. Which policies should be included will depend on the other critical factors. However, even without knowing those factors, we can already conclude that the set should not contain *dominated* policies, i.e., it should not contain any policies that are suboptimal for all scalarizations.

Definition 3.4   The *undominated set (U)* of an MODP is the set of all policies and associated value vectors that are optimal for some $\mathbf{w}$ of a scalarization function $f \in \mathcal{F}$:

$$U(\Pi) = \left\{ \mathbf{V}^{\pi} : \pi \in \Pi \ \wedge \ \exists f \in \mathcal{F} \ \exists \mathbf{w} \ \forall \pi' \in \Pi \ \ f(\mathbf{V}^{\pi}, \mathbf{w}) \geq f(\mathbf{V}^{\pi'}, \mathbf{w}) \right\},$$

where Π is the set of all allowed policies and $\mathcal{F}$ is the set of all scalarization functions that a user may have. For convenience, we assume that payoff vectors in $U(\Pi)$ contain both the value vectors and associated policies.

$U(\Pi)$ contains all policies that are optimal for some f and $\mathbf{w}$. Although this set contains no policies that are dominated, it may well contain *redundant* policies. In fact, we only need a set with at least *one* optimal policy for every f and $\mathbf{w}$. We call such a lossless subset of $U(\Pi)$ a *coverage set*, as it "covers" every f and $\mathbf{w}$ with an optimal policy.

Definition 3.5   A *coverage set (CS)*, $CS(\Pi)$, is a subset of $U(\Pi)$, such that for each possible $\mathbf{w}$, there is at least one optimal solution in the CS, i.e.,

$$\forall f \in \mathcal{F} \ \forall \mathbf{w} \ \exists \pi \ \left(\pi \in CS(\Pi) \wedge \forall \pi' \ f(\mathbf{V}^{\pi}, \mathbf{w}) \geq f(\mathbf{V}^{\pi'}, \mathbf{w}) \right).$$

Table 3.2: A simple example illustrating the difference between an undominated set $U(\Pi) = \{\pi_1, \pi_2, \pi_3\}$ and a coverage set $CS(\Pi)$, which could be $\{\pi_1, \pi_2 \ \pi_3\}$, $\{\pi_1, \pi_2\}$, or $\{\pi_2, \pi_3\}$

V_w^{π}	$w = \mathbf{true}$	$w = \mathbf{false}$
$\pi = \pi_1$	5	0
$\pi = \pi_2$	0	5
$\pi = \pi_3$	5	2
$\pi = \pi_4$	2	2

Note that a CS is not necessarily unique. Typically, we seek the smallest possible CS. Table 3.2 shows a simple example illustrating the difference between an undominated set and a coverage set. In this example, f is parameterized by only one binary weight feature $w \in \{false, true\}$. Note that this does not imply that there are two objectives. On the contrary, the number of objectives is not specified. Instead, it simply means that there are two possible scalarization functions, each corresponding to a different value of w. Π consists of four policies and the table shows the value of each policy under each scalarization. The undominated set $U(\Pi) = \{\pi_1, \pi_2, \pi_3\}$ because each of these policies has a maximal value of 5 for some w. While $U(\Pi)$ is also a coverage set

$CS(\Pi)$, it is not a minimal one. Since there are only two possible weights, it is possible to cover these weights with only two policies, one for each weight. Thus, $\{\pi_1, \pi_2\}$ and $\{\pi_2, \pi_3\}$ are also coverage sets, and are minimal.

Once we have computed a hopefully minimal coverage set, it can be used to speed up the selection of a policy during the selection phase (see Figure 1.1). For example, in the unknown weights scenario, the coverage set is computed during the planning phase, when the weights are unknown. However, the weights are revealed before the selection phase, enabling the selection of the policy with the maximal scalarized value for the given weights:

$$\pi^* = \underset{\pi \in CS(\Pi)}{\operatorname{argmax}} V_{\mathbf{w}}^{\pi}. \tag{3.4}$$

Note that performing the argmax over Π instead of $CS(\Pi)$ would yield the original scalarized optimization problem. Hence, multi-objective algorithms can be thought of as *pruning* algorithms that prune dominated policies from Π so as to simplify the selection of a particular policy during the selection phase.

In the decision support scenario, the weights are never explicitly revealed but the same interpretation nonetheless applies: the user selects a policy from $CS(\Pi)$, a process which typically gets easier as more policies from Π are pruned away.

In principle, it is possible that no policies can be pruned, and the smallest $CS(\Pi)$ is just Π. In this case, multi-objective methods offer no leverage and can even be harmful, as they may waste computational resources confirming that no policies can be pruned. However, in practice, there are typically coverage sets that are much smaller than Π, such that efficient multi-objective methods can offer great leverage.

3.2.1   CASE #1: LINEAR SCALARIZATION AND A SINGLE POLICY

We now proceed to discuss the appropriate solution concept for each of the six cases shown in Table 3.1, starting with the top-left box labeled (1). This case assumes a linear scalarization function and a single policy (i.e., known weights); whether the policies must be deterministic or not is irrelevant for this case.

It turns out that, in this case, the presence of multiple objectives poses no difficulties. Consider the case of an MOMDP: because the inner product computed by f distributes over addition, it can be pushed inside the expectation that appears in the definition of the value function (2.3). In the infinite horizon setting this leads to:

$$V_{\mathbf{w}}^{\pi} = \mathbf{w} \cdot \mathbf{V}^{\pi} = \mathbf{w} \cdot E\left[\sum_{k=0}^{\infty} \gamma^k \mathbf{r}_{t+k+1}\right] = E\left[\sum_{k=0}^{\infty} \gamma^k (\mathbf{w} \cdot \mathbf{r}_{t+k+1})\right]. \tag{3.5}$$

Intuitively, this is the value function of π in a new single-objective MDP that at time t generates a scalar reward $\mathbf{w} \cdot \mathbf{r}_t$. Since this single-objective MDP has additive returns, it can be solved with standard methods, yielding a single policy. Due to Theorem 2.7, a deterministic stationary policy suffices.

Note that this case is the only one in the taxonomy for which multi-objective methods are not required. However, even in this case, a multi-objective approach can still be preferable, e.g., $\mathbf{V}^\pi$ may be easier to estimate than $V_\mathbf{w}^\pi$ in large or continuous MOMDPs where function approximation is required (see Chapter 6).

3.2.2   CASE #2: LINEAR SCALARIZATION AND MULTIPLE POLICIES

We now turn to the top-right box of Table 3.1, labeled (2), which assumes a linear scalarization function but multiple policies (i.e., unknown weights or decision support); as before, whether the policies must be deterministic or not is irrelevant.

Since we do not know $\mathbf{w}$ during planning or learning, we want to find a coverage set. If f is linear, then $U(\Pi)$, which is automatically a coverage set, consists of the *convex hull*.[1] Substituting (3.1) in the definition of the undominated set (Definition 3.4) yields the definition of the convex hull:

Definition 3.6   The *convex hull (CH)* is the undominated set for non-decreasing linear scalarizations $f(\mathbf{V}^\pi, \mathbf{w}) = \mathbf{w} \cdot \mathbf{V}^\pi$:

$$CH(\Pi) = \left\{ \mathbf{V}^\pi : \pi \in \Pi \ \wedge \ \exists \mathbf{w} \forall \pi' \ \ \mathbf{w} \cdot \mathbf{V}^\pi \geq \mathbf{w} \cdot \mathbf{V}^{\pi'} \right\},$$

where $\mathbf{w}$ adheres to the simplex constraints, i.e., $\forall i \ \ w_i \geq 0$ and $\sum_i w_i = 1$.

Figure 3.1 (left) illustrates the concept of a convex hull. Each point in the plot represents the multi-objective value of a given policy for a two-objective problem. The axes represent the reward dimensions. The convex hull consists of the red and blue points, connected by lines that form a convex surface. We refer to this type of visualization as the *value-space perspective*.

Given a linear f, the scalarized value of each policy is a linear function of the weights. This is illustrated in Figure 3.1 (right), where the x-axis represents the weight for dimension 0 ($\mathbf{w}[1] = 1 - \mathbf{w}[0]$), and the y-axis the scalarized value of the policies. We refer to this type of visualization as the *weight-space perspective*. To select a policy for a given $\mathbf{w}$, we need only know the values of the convex hull policies, which form the *upper surface* of the scalarized value, as illustrated by the red and blue lines, and correspond to the four convex hull policies in Figure 3.1 (left). The upper surface forms a piecewise linear and convex function. Such functions are also well-known from the literature on partially-observable Markov decision processes (POMDPs) [Kaelbling et al., 1998], whose relationship to MOMDPs we discuss in Section 5.2.1.

Like any $U(\Pi)$, $CH(\Pi)$ can contain superfluous policies. However, we can also define the *convex coverage set (CCS)* as the specification of the coverage set when f is linear. A CCS retains at least one policy from the CH that maximizes the scalarized payoff, $\mathbf{w} \cdot \mathbf{V}^\pi$, for every $\mathbf{w}$. We can

---

[1]The term *convex hull* is overloaded. In geometry (e.g., Jarvis [1973]), the convex hull is a superset of what we mean by the convex hull in this book.

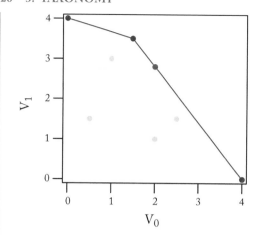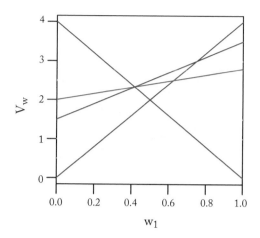

Figure 3.1: Example of the convex hull and convex coverage set. Each point at left represents the multi-objective value of a given policy. We refer to the type of visualization on the left as the *value-space perspective*. Each line at right represents the linearly scalarized value of a policy across values of **w**. We refer to this type of visualization as the *weight-space perspective*. The convex hull consists of the three red points and the blue point. However, a convex coverage set can be formed using only the red points because the blue point's corresponding vector at right contributes to the upper surface but is not necessary to represent it.

obtain the definition of the CCS by substituting Equation 3.1 in the definition of the coverage set (Definition 3.5):

Definition 3.7 A *convex coverage set (CCS)*, $CCS(\Pi) \subseteq CH(\Pi)$, is a CS for linear non-decreasing scalarizations, i.e.,

$$\forall \mathbf{w} \exists \pi \; \left(\pi \in CCS(\Pi) \wedge \forall \pi' \; \mathbf{w} \cdot \mathbf{V}^{\pi} \geq \mathbf{w} \cdot \mathbf{V}^{\pi'} \right).$$

Figure 3.1 (left) also illustrates the difference between the CH and a CCS. In particular, a CCS can be constructed using only the red points, excluding the blue one. Figure 3.1 (right) shows why: while the blue vector (corresponding to the blue point) contributes to the upper surface, it is possible to represent that surface without the blue vector (using only the red vectors) as the blue vector is undominated at only one point and other vectors are undominated there too.

For deterministic stationary policies, the difference between $CH(\Pi)$ and $CCS(\Pi)$ may often be small. Therefore, the terms are often used interchangeably. However, in the case of non-stationary or stochastic policies, the difference is quite significant, as the CH can contain infinitely many policies, while it is possible to construct a finite CCS.

Fortunately, the following result, which is analogous to Theorem 2.7 but for MOMDPs, shows that we can restrict our attention to deterministic stationary policies.

Corollary 3.8   *For any MOMDP, a $CCS(\Pi_{DS})$ is also a $CCS(\Pi)$.*

Proof. If f is linear, we can translate the MOMDP to a single-objective MDP, for each possible $\mathbf{w}$. This is done by treating the inner product of the reward vector and $\mathbf{w}$ as the new rewards, and leaving the rest of the problem as is. Since the inner product distributes over addition, the scalarized returns remain additive (3.5). Thus, for every $\mathbf{w}$ there exists a translation to a single-objective MDP, for which an optimal deterministic and stationary policy must exist, due to Theorem 2.7. Hence, for each $\mathbf{w}$ there exists an optimal deterministic stationary policy. Therefore, there exists a $\pi \in CCS(\Pi_{DS})$ that is optimal for that $\mathbf{w}$. Consequently, there cannot exist a $\pi' \in \Pi \setminus \Pi_{DS}$ such that $\mathbf{w} \cdot \mathbf{V}^{\pi'} > \mathbf{w} \cdot \mathbf{V}^{\pi}$ and thus $CCS(\Pi_{DS})$ is also a $CCS(\Pi)$. $\square$

3.2.3   CASE #3: MONOTONICALLY INCREASING SCALARIZATION AND A SINGLE DETERMINISTIC POLICY

We now turn to cases where the scalarization is not necessarily linear but instead can only be assumed to be monotonically increasing. We start with box (3) in Table 3.1, the case for which a single deterministic policy is required.

The question then arises: Can we restrict our attention to stationary policies, as in Case #1? In that setting, we were able to push the scalarization inside the expectation computed by the value function (see (3.5)) and then use Theorem 2.7 to show that deterministic stationary policies suffice.

Unfortunately, this trick no longer works if f is not known to be linear, because nonlinear scalarization and expectation do not commute:

$$V_{\mathbf{w}}^{\pi} = f(\mathbf{V}^{\pi}, \mathbf{w}) = f\left(E\left[\sum_{k=0}^{\infty} \gamma^k \mathbf{r}_{t+k+1}\right], \mathbf{w}\right) \neq E\left[\sum_{k=0}^{\infty} \gamma^k f(\mathbf{r}_{t+k+1}, \mathbf{w})\right]. \tag{3.6}$$

Consequently, the scalarized value can no longer be written as a sum, i.e., nonlinear scalarization destroys the additivity property. We can thus no longer appeal to Theorem 2.7 to justify ignoring non-stationary policies. In fact, the following theorem shows that, for this setting, non-stationary policies may be better than the best stationary ones.

Theorem 3.9   *In infinite-horizon MOMDPs, deterministic non-stationary policies can Pareto-dominate deterministic stationary policies that are undominated by other deterministic stationary policies [White, 1982].*

To see why, consider the following MOMDP adapted from an example by White [1982]. There is only one state and three actions a_1, a_2, and a_3, which yield rewards $(3, 0)$, $(0, 3)$, and $(1, 1)$, respectively. If we allow only deterministic stationary policies, then there are three possible policies $\pi_1, \pi_2, \pi_3 \in \Pi_{DS}$, each corresponding to always taking one of the actions, none of which Pareto-dominate each other. These policies have the following state-independent values:

$$\mathbf{V}^{\pi_1} = (3/(1-\gamma), 0), \tag{3.7}$$
$$\mathbf{V}^{\pi_2} = (0, 3/(1-\gamma)), \tag{3.8}$$
$$\mathbf{V}^{\pi_3} = (1/(1-\gamma), 1/(1-\gamma)). \tag{3.9}$$

However, if we now consider the set of possibly non-stationary policies Π_D, we can construct a policy $\pi_{ns} \in \Pi_D \setminus \Pi_{DS}$ that alternates between a_1 and a_2, starting with a_1, and whose value is:

$$\mathbf{V}^{\pi_{ns}} = (3/(1-\gamma^2), 3\gamma/(1-\gamma^2)). \tag{3.10}$$

Consequently, $\pi_{ns} \succ_P \pi_3$ when $\gamma \geq 0.5$ and thus we cannot restrict our attention to stationary policies.[2] Figure 3.2 shows the multi-objective value of each policy for different values of γ.

3.2.4   CASE #4: MONOTONICALLY INCREASING SCALARIZATION AND A SINGLE STOCHASTIC POLICY

The next case we consider is exactly the same as the one above save that stochastic policies are permitted: the scalarization function remains potentially nonlinear and a single policy is sought.

In this case, we again cannot consider only deterministic stationary policies. However, we can employ stochastic stationary policies instead of deterministic non-stationary ones. In particular, we can employ a *mixture policy* [Vamplew et al., 2009] π_m that takes a set of N deterministic policies, and selects the i-th policy from this set, π_i with probability p_i, where $\sum_{i=0}^{N} p_i = 1$. This leads to values that are a linear combination of the values of the constituent policies. In our previous example, we can replace π_{ns} by a policy π_m that chooses π_1 with probability p_1 and π_2 otherwise, resulting in the following values:

$$\mathbf{V}^{\pi_m} = p_1 \mathbf{V}^{\pi_1} + (1-p_1)\mathbf{V}^{\pi_2} = \left(\frac{3p_1}{1-\gamma}, \frac{3(1-p_1)}{1-\gamma}\right).$$

If $\frac{1}{3} \leq p_1 \leq \frac{2}{3}$, then $\pi_m \succ_P \pi_3$. Hence, a suitable solution for this setting is a mixture policy of two or more deterministic stationary policies.[3]

3.2.5   CASE #5: MONOTONICALLY INCREASING SCALARIZATION AND MULTIPLE DETERMINISTIC POLICIES

We now consider cases where the scalarization function is only assumed to be monotonically increasing but where we seek multiple policies, starting with the deterministic case.

---

[2]White [1982] shows this in an infinite-horizon discounted setting, but the arguments hold also for the finite-horizon and average-reward settings.

[3]A mixture policy is both stochastic and non-stationary, as it conditions on something other than the state when picking its action: the outcome of the random event that determines which of the constituent policies to execute. In theory, it is also possible to require stationary policies without requiring deterministic policies [Clempner, 2016].

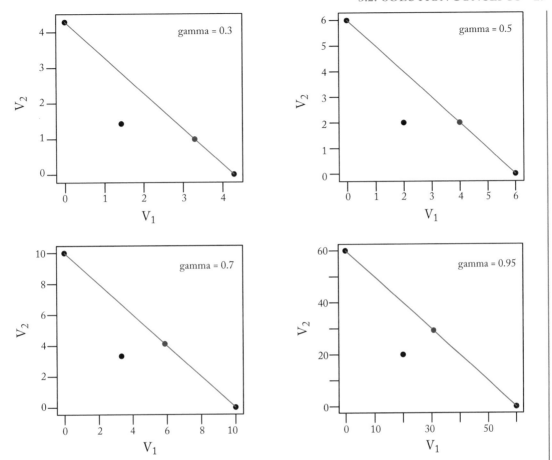

Figure 3.2: The multi-objective value of the deterministic stationary policies (black dots) in White's example and the deterministic non-stationary policy (red dot) for different values of γ.

Just as we specialized the undominated set to the case of linear scalarization functions, yielding the convex hull, we can also specialize the undominated set to the case of monotonically increasing scalarization functions, yielding the *Pareto front*.

Definition 3.10   The *Pareto front* (PF) is the set of all policies that are not Pareto-dominated:

$$ PF(\Pi) = \left\{ \pi \, : \, \pi \in \Pi \, \wedge \, \neg \exists \pi' \in \Pi \, \mathbf{V}^{\pi'} \succ_P \mathbf{V}^\pi \right\}. $$

Figure 3.3 uses the same scenario depicted in Figure 3.1 to illustrate an example of a Pareto front. As before, the red and blue points together comprise the CH. Now, the addition of the black point yields a Pareto front. The black point belongs in the PF because no other point Pareto-dominates it. This can be seen by noting that the region to the upper right of the black point (marked with the dashed lines) contains no points. The black point does not belong in the CH because it does not contribute to the undominated upper surface shown at right.

A point can be Pareto optimal while not contributing to this upper surface because Pareto dominance is pairwise: a single policy must be found that is superior for all weights. In the example, for every $\mathbf{w}$, there exists a vector with higher scalarized value than the black vector, which is why it is not in the CH, but there is no single vector that does so for all $\mathbf{w}$, which is why it is in the PF.

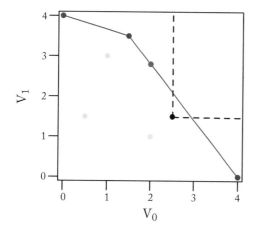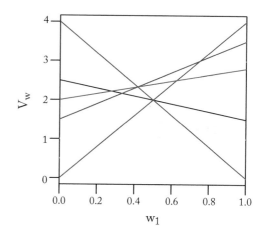

Figure 3.3: Example of the Pareto front. Each point at left represents the multi-objective value of a given policy and each line at right represents the linearly scalarized value of a policy across values of $\mathbf{w}$. A minimal CCS can be made from the three red points. The addition of the blue point forms the CH. The addition of the black point forms the Pareto front. Since no point is to the upper right of the black point (in the region marked with the dashed lines), it is not Pareto-dominated. However, since the figure at right shows that it does not contribute to the undominated upper surface, it is not in the convex hull.

Note that $PF(\Pi)$ is not the set of undominated policies $U(\Pi)$ for all specific strictly monotonically increasing f. We have already seen that for the special case of linear f, $U(\Pi) = CH(\Pi)$, which is a subset of $PF(\Pi)$. However, for any strictly monotonically increasing f, we know that a policy that is not in the $PF(\Pi)$ is dominated with respect to f, i.e., $\pi \notin PF(\Pi) \Rightarrow \pi \notin U(\Pi)$. This is because, for strictly monotonically increasing f and $\pi \notin PF(\Pi)$, there cannot exist a $\mathbf{w}$ for which π is optimal, since by definition there exists a π' such that $\mathbf{V}^{\pi'} \succ_P \mathbf{V}^{\pi}$ and, since f is strictly monotonically increasing, this implies that $V_{\mathbf{w}}^{\pi'} > V_{\mathbf{w}}^{\pi}$.

However, if we know only that f is strictly monotonically increasing, we cannot settle for a subset of $PF(\Pi)$ either, because there exist strictly monotonically increasing f for which

$U(\Pi^m) = PF(\Pi^m)$. Perny and Weng [2010] show that $U(\Pi) = PF(\Pi)$ for the *Tchebycheff function*, which is strictly monotonically increasing. Therefore, we cannot discard policies from the $PF(\Pi)$ and retain an undominated set $U(\Pi)$ for all strictly monotonically increasing f.

Like any undominated set, $PF(\Pi)$ may contain redundant policies. Therefore, we define the *Pareto coverage set* (PCS), a special case of the coverage set for monotonically increasing scalarization functions:

Definition 3.11   A *Pareto coverage set (PCS)*, $PCS(\Pi) \subseteq PF(\Pi)$, is a lossless subset of $PF(\Pi)$, i.e., it only needs to contain each unique value-vector in the PF once:

$$\mathbf{V}^\pi = \mathbf{V}^{\pi'} \implies \pi \in PCS(\Pi) \vee \pi' \in PCS(\Pi) \vee \pi \notin PF(\Pi).$$

Again, for deterministic stationary policies, the difference between a $PCS(\Pi)$ and $PF(\Pi)$ may be minor. Note that $PF(\Pi)$ is automatically a $PCS(\Pi)$. Most papers in the literature therefore take $PF(\Pi)$ as the solution concept.

3.2.6   CASE #6: MONOTONICALLY INCREASING SCALARIZATION AND MULTIPLE STOCHASTIC POLICIES

We now consider the sixth and final case: monotonically increasing scalarization functions with multiple stochastic policies. A natural solution concept for this case is a PCS of stochastic stationary policies. As in the single policy settings, a monotonically increasing scalarization function that might not be linear implies that we cannot consider only stationary deterministic policies. However, if stochastic policies are allowed, we do not need to consider non-stationary policies. The result is a PCS of stochastic stationary policies.

However, because this set is in general not finite, it poses practical problems for any algorithm that hopes to compute it. Fortunately, it is not necessary to explicitly represent an entire $PCS(\Pi)$. Instead, it is sufficient to compute a $CCS(\Pi_{DS})$, i.e., a convex coverage set of deterministic stationary policies. The necessary stochastic policies to create a $PCS(\Pi)$ can then be easily constructed by making mixture policies, as in Section 3.2.4, from those policies in the $CCS(\Pi_{DS})$.

Corollary 3.12   *In an infinite horizon discounted MOMDP, an infinite set of mixture policies P_M can be constructed from policies that are on a $CCS(\Pi_{DS})$, such that this set P_M, is a $PCS(\Pi)$ [Vamplew et al., 2009].*

Proof. We can construct a policy with any value vector on the convex surface, e.g., the red lines in Figure 3.3 (left), by mixing policies on a $CCS(\Pi_{DS})$, e.g., the red dots.[4] Therefore, we can always

---

[4]We should always mix policies that are "adjacent"; the line between any pair of the policies we mix should be on the convex surface. For example, mixing the policy represented by the leftmost red dot in Figure 3.3 (left) and the policy represented by the rightmost red dot does not lead to optimal policies, as the line connecting these two points is under the convex surface.

construct a mixture policy that dominates a policy with a value under this surface, such as the black dot. We can show by contradiction that there cannot be any policy above the convex surface. If there was, it would be optimal for some **w** if f was linear. Consequently, by Corollary 3.8, there would be a deterministic stationary policy with at least equal value. But since the convex surface spans the values on the $CCS(\Pi_{DS})$, this leads to a contradiction. Thus, no policy can Pareto-dominate a mixture policy on the convex surface. □

Thanks to Corollary 3.12, it is sufficient to compute a $CCS(\Pi_{DS})$ to solve an MOMDP. A surprising consequence of this fact, which to our knowledge is not made explicit in the literature, is that Pareto optimality, though the most common solution concept associated with multi-objective problems, is actually only necessary in one specific problem setting:

Observation 3.13   The multiple policy setting when f is monotonically increasing and only deterministic policies are considered (box (5) of Table 3.1) requires computing a Pareto coverage set. When either f is linear or stochastic policies are allowed, a $CCS(\Pi_{DS})$ suffices.

Wakuta [1999] proves the sufficiency of a $CCS(\Pi_{DS})$ for monotonically increasing scalarizations with multiple stochastic policies in infinite horizon MOMDPs, but using a different approach than Corollary 3.12. Instead of mixture policies, he uses stationary randomizations over deterministic stationary policies. Wakuta and Togawa [1998] provide a similar proof for the average reward case.

Note that, while it is common to consider non-stationary or stochastic policies when f is nonlinear, such policies typically condition only on the current state, or the current state and time, not the agent's reward history. However, in this setting, policies that condition on that reward history can dominate those that do not. For example, suppose there are two objectives which can take only positive values and f simply selects the smaller of the two, i.e., $f(\mathbf{V}^\pi, \mathbf{w}) = \min_i V_i^\pi$. Suppose also that, in a given state, two actions are available, which yields rewards of $(4, 4)$ and $(0, 5)$ respectively. Finally, suppose that the agent can arrive at that state with one of two reward histories, whose discounted sums are either $(5, 0)$ or $(3, 3)$. A policy that conditions on these discounted reward histories can outperform policies that do not, i.e., the optimal policy selects the action yielding $(4, 4)$ when the reward history sums to $(3, 3)$ and the action yielding $(0, 5)$ when the reward history sums to $(5, 0)$. So, while for single objective MDPs the Markov property and additive returns are sufficient to restrict our attention to policies that ignore history, in the multi-objective case, the scalarized returns are no longer additive and therefore the optimal policy can depend on the history.

3.3   IMPLICATIONS FOR MO-COGS

The preceding discussion of the taxonomy focused on MOMDPs. In this section, we briefly discuss the implications for MO-CoGs. Overall, the situation is quite similar to that of MOMDPs. The main difference is that there is no notion of a non-stationary policy, since a MO-CoG is not a sequential problem.

In particular, given known weights and a linear scalarization function (box 1), we again do not require special methods, as the MO-CoG can be converted to a CoG by scalarizing each local payoff function, i.e., we can move $\mathbf{w}$ inside the expectation, just as in (3.5) but where the inner summation goes over the local payoff functions instead of timesteps:

$$u_{\mathbf{w}}^{\pi} = \mathbf{w} \cdot \mathbf{u}^{\pi} = \mathbf{w} \cdot E\left[\sum_{k=1}^{\rho} \mathbf{u}^k(\mathbf{a})\right] = E\left[\sum_{k=1}^{\rho} \mathbf{w} \cdot \mathbf{u}^k(\mathbf{a})\right], \tag{3.11}$$

where π specifies a distribution over the joint action $\mathbf{a}$. As with any CoG, a deterministic policy suffices. If the weights are not known (box 2), a $CCS(\Pi_D)$ suffices.

Given known weights but a monotonically increasing scalarization function, then a stochastic policy may be preferable, for the same reasons as in an MOMDP. If stochastic policies are allowed (box 4), we can construct one using mixture policies. If deterministic policies are required (box 3), then we cannot mimic a mixture policy with a non-stationary one, and thus the best deterministic policy may have less value than the best stochastic policy.

If the weights are not known, then a similar logic applies. If stochastic policies are permitted (box 6), then a $CCS(\Pi_D)$ suffices as it can be used to construct the needed mixture policies. Otherwise (box 5), we need a $PCS(\Pi_D)$. For some weights, the best policy in that $PCS(\Pi_D)$ may be inferior to some mixture policies that could be formed from the $CCS(\Pi_D)$, were that permitted.

3.4 APPROXIMATE SOLUTION CONCEPTS

So far, we have only discussed *exact* solution concepts, e.g., a coverage set contains an *optimal* policy for each $\mathbf{w}$. However, in many settings, computing such exact solutions may be infeasible. Therefore, in this section, we discuss approximate variants of the PCS and CCS. We start with a helpful definition:

Definition 3.14   For a given solution set S and some family of scalarization functions $\mathcal{F}$, the maximum utility loss $MUL(S, \mathcal{F})$ is the maximum scalarized value that is lost due to approximation:

$$\forall f \in \mathcal{F} \ \forall \mathbf{w} \ \forall \mathbf{V}^{\pi} \in CS(\Pi) \ \exists \mathbf{V}^{\pi'} \in S \quad f(\mathbf{V}^{\pi}, \mathbf{w}) \leq f(\mathbf{V}^{\pi'}, \mathbf{w}) + MUL(S, \mathcal{F}),$$

where $CS(\Pi)$ is the coverage set appropriate for $\mathcal{F}$.

Several approximate variants of the PCS have been proposed. One of the most popular is the ε-PCS [Zitzler et al., 2003].[5] There are multiple definitions of the ε-PCS; here, we provide the definition of the *additive ε-PCS*. There is also a *multiplicative ε-PCS* [Zintgraf et al., 2015].

Definition 3.15 A given solution set S is an additive ε-PCS if

$$\forall \mathbf{V}^\pi \in PCS(\Pi) \ \exists \mathbf{V}^{\pi'} \in S \ : \ \forall i = 1, \ldots, d \ : \ V_i^\pi \leq V_i^{\pi'} + \varepsilon,$$

where d is the number of objectives.

Note that an ε-PCS may not contain any undominated solutions—$S \cap PF(\Pi)$ may be an empty set—but at least the maximal difference between a value vector in the PCS and the closest value vector in the ε-PCS is at most ε in all dimensions.

However, a fundamental limitation of an ε-PCS for arbitrary monotonically increasing scalarizations is that *any* increase in any objective may lead to an infinite increase in user utility. Therefore, it is impossible to compute the MUL of an ε-PCS without more information about f and $\mathbf{w}$. For example, if we know that f is Lipschitz-continuous with a Lipschitz-constant L, the MUL is bounded by $\varepsilon \sqrt{d} L$ [Zintgraf et al., 2015].

For linear scalarization functions, we have much more information, i.e., we know the exact shape of f, and that $\mathbf{w}$ adheres to the simplex constraints. In this case, we can formulate an ε-CCS for which the MUL is ε.

Definition 3.16 A given solution set S is an ε-CCS if

$$\forall \mathbf{w} \quad \max_{\mathbf{V}^\pi \in CCS(\Pi)} \mathbf{w} \cdot V_i^\pi - \max_{\mathbf{V}^{\pi'} \in S} \mathbf{w} \cdot V_i^{\pi'} \leq \varepsilon,$$

where $\mathbf{w}$ is a linear weight vector adhering to the simplex constraints.[6]
Note that an ε-PCS is automatically an ε-CCS, though it need not be a minimal one.

3.5 BEYOND THE TAXONOMY

While the taxonomy presented in this chapter aims to cover as much research on multi-objective decision making as possible, it is nonetheless not completely comprehensive. In this section, we briefly mention a few formalisms that fall outside the scope of our taxonomy.

In particular, our taxonomy, and also the entire utility-based perspective, assumes the existence of a scalarization function f. Furthermore, we consider only monotonically increasing scalarization functions. We argue that this is a minimal assumption because it simply states that increasing the value with respect to one objective without diminishing the others cannot decrease the scalarized value. This corresponds well with what an objective is commonly understood to be.

---

[5]The ε-PCS is called ε-approximate Pareto front by Zitzler et al. [2003]. We use ε-PCS for consistency with our own terminology.

[6]We can always make $\mathbf{w}$ adhere to the simplex constraints by dividing by a constant c. If $\mathbf{w}$ is not on the simplex, the MUL reported here should be multiplied by this c.

However, it is possible to have a preference ordering over value vectors to which no f corresponds. For example, when preferences are based on a lexicographical ordering, i.e., even an infinitesimal improvement in an objective with a higher priority is better than a large improvement in an objective with a lower priority, and the rewards are vectors in $\mathfrak{R}^n$, there is no corresponding $f : \mathfrak{R}^n \to \mathfrak{R}$ [Sen, 1995]. Therefore, lexicographical orderings fall outside of our mathematical formalism.

Furthermore, our taxonomy considers only one special case of the monotonically increasing scalarization function, namely linear scalarization functions. However, other special cases have also been considered. For example, certain fairness constraints on f lead to the *Lorenz front* [Perny et al., 2013], a subset of the Pareto front.

CHAPTER 4

Inner Loop Planning

Most algorithms for solving multi-objective problems are constructed by extending algorithms for the corresponding single-objective problem. There are two main approaches for constructing such extensions: the *inner loop* and *outer loop* approaches. In the inner loop approach, which we discuss in this chapter, the inner workings of the single-objective algorithm are adapted by exchanging sums and maximizations by cross-sums and pruning operators. In the *outer loop approach*, which we discuss in the next chapter, the single-objective algorithm is left intact, but a wrapper, i.e., an outer loop, is placed around it.

4.1   INNER LOOP APPROACH

A key difference between SODPs and MODPs is that a solution to the former is a single policy that maximizes a scalar value, while in the latter it is typically a set of policies and associated value vectors, i.e., a coverage set.

In the inner loop approach, an algorithm for computing such a coverage set is created by changing the operators required for optimization in single-objective problems to ones that work on sets, and subsequently *pruning* away those vectors that are not needed to form a CS. Using these multi-objective operators, the inner loop solves a series of smaller multi-objective problems. These smaller problems are called *local* problems because they involve only some parts of the original problem. The key insight behind the inner loop approach is that if these local problems are chosen and conditioned well, then their solutions, i.e., local CSs, can directly lead to a full CS. For example, in MO-CoGs, the local subproblems typically consist of a few local payoff functions, and are conditioned on the actions that are directly adjacent to the local subproblem, as we discuss in detail in Section 4.2.

4.1.1   A SIMPLE MO-COG

To illustrate how the inner loop approach works, we first describe a simple MO-CoG that can easily be solved. Starting from the corresponding single-objective algorithm, we provide a step-by-step approach to creating an inner loop multi-objective algorithm.

The MO-CoG consists of two local payoff functions: called $\mathbf{u}_l(a_l)$ and $\mathbf{u}_r(a_r)$ with for each local action (A, B, C, or D), an expected reward as given in Table 4.1. Note that the scopes of these local reward functions do not overlap, i.e., they do not depend on the same local actions. This complete independence between the local payoff functions makes it easy to solve. We show how to solve MO-CoGs in general in Section 4.2.

In the simple MO-CoG of Table 4.1, a deterministic policy takes one action a_l and one action a_r. The value of a deterministic policy is the sum of the local rewards:

$$\mathbf{u}(\mathbf{a}) = \mathbf{u}_l(a_l) + \mathbf{u}_r(a_r).$$

For example, for the deterministic policy $\mathbf{a} = (A_l, A_r)$, the value vector is $\mathbf{u}(A_l, A_r) = (5.7, 6.9) + (7.3, 7.6) = (13, 14.5)$.

Table 4.1: A simple MODP: select one element from each list and receive the associated reward vector

A_l	$(5.7, 6.9)$		A_r	$(7.3, 7.6)$
B_l	$(7.1, 5.7)$		B_r	$(5.9, 8.2)$
C_l	$(7.5, 5.4)$		C_r	$(8.8, 6.4)$
D_l	$(6.6, 6.7)$		D_r	$(6.6, 7.7)$
$\mathbf{u}_l(a_l)$			$\mathbf{u}_r(a_r)$	

In the corresponding single-objective version of this problem, the value would be scalar,

$$u(\mathbf{a}) = u_l(a_l) + u_r(a_r),$$

and the optimal value could be found by maximization:

$$\max_{\mathbf{a}} u(\mathbf{a}) = \max_{\mathbf{a}}(u_l(a_l) + u_r(a_r)) = \max_{a_l} u_l(a_l) + \max_{a_r} u_r(a_r). \tag{4.1}$$

However, in the multi-objective setting, we cannot maximize because the utilities are vector-valued and thus a maximizing action need not exist. Therefore, in the inner loop approach, we first reformulate the problem in terms of sets and then replace the maximization and summation operators with *pruning* and *cross-sum* operators.

To reformulate the problem in terms of sets, we turn $\mathbf{u}_l$ and $\mathbf{u}_r$ into sets of value vectors rather than reward functions:

$$\mathcal{U}_l = \{\mathbf{u}_l(a_l) : a_l \in \{A_l, B_l, C_l, D_l\}\},$$

and accordingly for $\mathbf{u}_r$. For convenience, we assume that $\mathcal{U}_l$ and $\mathcal{U}_r$ contain both the value vectors and the local actions.

Then, we replace the maximization operator with a *pruning* operator that removes dominated solutions, yielding a coverage set. Computing a CS can be seen as computing the maximal value *for every possible scalarization in parallel*. The pruning operator takes a set of vectors, such as $\mathcal{U}_l$, and removes all vectors that do not maximize the value for any weight $\mathbf{w}$ of the scalarization function f. In other words, we first solve the local subproblem for $\mathcal{U}_l$, resulting in a local CS. Note that the original maximization operator can be seen as a degenerate form of pruning (removing all vectors except the maximizing ones) that is appropriate when $\mathbf{w}$ is known, i.e., in a single-objective setting.

When computing a PCS, we need a different pruning operator than when computing a CCS. In its abstract form, i.e., when the operator computes a local CS (either a CCS or PCS), we denote this operator as LCS.

Next, we replace the summation operator with a *cross-sum* operator, $\oplus$, which combines two sets by computing the sum of all possible pairs with one element from each set:

$$\mathcal{A} \oplus \mathcal{B} = \{\mathbf{A} + \mathbf{B} : \mathbf{A} \in \mathcal{A} \wedge \mathbf{B} \in \mathcal{B}\}. \tag{4.2}$$

Finally, we use these new operators to translate (4.1) into a multi-objective optimization problem, yielding an inner loop method for the problem described in Table 4.1:

$$CS = LCS(LCS(\mathcal{U}_l) \oplus LCS(\mathcal{U}_r)). \tag{4.3}$$

Note that while there are only two maximization operations in (4.1), there are three LCS operations in (4.3). The outermost LCS operation is necessary because the combination of two sets may contain combinations that are dominated. By contrast, the sum of maximal local rewards in the single-objective problem form a single value when summed.

We have now defined the general schema for creating a multi-objective inner loop method from a base single-objective method: transform the original function-based problem to a set-based problem, replace the maximizations by pruning operations, and replace the summations by cross-sums. Before we can use such a method in practice, however, we must:

(a) implement LCS such that it yields the appropriate solution concept,

(b) select the places in the algorithm where pruning is applied, and

(c) confirm that the resulting inner loop method correctly outputs the CS. In particular, we verify that no *necessary* vectors for computing the CS are pruned prematurely, and that no excess vectors, i.e., vectors that are never optimal for any allowed scalarization, are retained in the output.

We start by addressing (a). In this book, we treat two cases: the PCS and the CCS.

4.1.2   FINDING A PCS

Algorithm 4.1 shows PPrune, an implementation of LCS that yields a local PCS. PPrune computes a PCS by performing pairwise comparisons.

In each iteration, a P-undominated vector from the input $\mathcal{V}$ is identified. Starting from a random value vector $\mathbf{V}$, on line 3, pairwise comparisons are made with the other value vectors in $\mathcal{V}$. If $\mathbf{V}$ is Pareto-dominated by a vector $\mathbf{V}$, that vector replaces $\mathbf{V}$ as the current value vector (line 6). This process ensures that by the end of the comparisons on line 9, the final $\mathbf{V}$ is P-undominated

---

Algorithm 4.1 PPrune($\mathcal{V}$)

Input : A set of value vectors $\mathcal{V}$
Output : A PCS $\mathcal{V}^*$

1: $\mathcal{V}^* \leftarrow \emptyset$
2: **while** $\mathcal{V} \neq \emptyset$ **do**
3:      **V** $\leftarrow$ the first element of $\mathcal{V}$
4:      **for** each **V**$' \in \mathcal{V}$ **do**
5:          **if** **V**$' >_P$ **V** **then**
6:              **V** $\leftarrow$ **V**$'$
7:          **end if**
8:      **end for**
9:      Remove **V** , and all vectors (weakly) P-dominated by **V**, from $\mathcal{V}$
10:     Add **V** to $\mathcal{V}^*$
11: **end while**
12: **return** $\mathcal{V}^*$

---

with respect to the rest of the value-vectors in $\mathcal{V}$. After this **V** has been identified, it is moved to the result set $\mathcal{V}^*$ that will become a full PCS. Furthermore, all vectors dominated by **V** are removed from $\mathcal{V}$. This removal ensures that no vectors in $\mathcal{V}$ are dominated by one in $\mathcal{V}^*$. In other words, no excess value vectors are retained. Therefore, in subsequent iterations, we know that no vector dominated by a vector in $\mathcal{V}^*$ is added to $\mathcal{V}^*$.

Theorem 4.1   *The computational complexity of* PPrune *as defined by Algorithm 4.1 is*

$$O(d|\mathcal{V}||PCS|), \qquad (4.4)$$

where $|\mathcal{V}|$ is the size of the input, $|PCS|$ the size of the output PCS, and the number of objectives is d. This runtime is achieved because each iteration finds one PCS value vector, yielding $|PCS|$ iterations. In each iteration, at most $|\mathcal{V}|$ pairwise comparisons are made between two vectors of length d.

To address (b), let us implement the algorithm for our simple example problem as:

$$PCS = \text{PPrune}(\text{PPrune}(\mathcal{U}_l) \oplus \text{PPrune}(\mathcal{U}_r)).$$

To address (c), we can confirm that this algorithm is correct by showing 1) that no excess (i.e., P-dominated) value vectors are retained and 2) that no necessary vectors (i.e., P-undominated) are removed. First, note that any vector in the result is a summation between a vector from $\mathcal{U}_l$ and a vector from $\mathcal{U}_r$. Because a subset of all possible such combinations is retained after PPrune($\mathcal{U}_l$) $\oplus$

PPrune($\mathcal{U}_r$) (as no new vectors are introduced by applying PPrune), and then PPrune is applied again, we know that 1) must be true, as PPrune retains no P-dominated vectors. Furthermore, we observe that:

$$\forall \mathbf{U} \quad \mathbf{V} \succ_P \mathbf{V}' \implies \mathbf{V} + \mathbf{U} \succ_P \mathbf{V}' + \mathbf{U},$$

i.e., if we have the sum of two vectors $\mathbf{V}' + \mathbf{U}$, and we replace $\mathbf{V}'$, by another vector $\mathbf{V}$, which dominates $\mathbf{V}'$, the resulting new sum P-dominates the original sum. Therefore, PPrune cannot delete necessary components of undominated sums from $\mathcal{U}_l$ and $\mathcal{U}_r$, and the algorithm must thus be correct.

4.1.3   FINDING A CCS

Algorithm 4.2 shows CPrune, an implementation of LCS that yields a local CCS. CPrune is based on an algorithm for CCS pruning [Feng and Zilberstein, 2004][1] with one modification. In order to improve runtime guarantees, CPrune first pre-prunes the input set, $\mathcal{V}$, to a PCS using the PPrune algorithm (Algorithm 4.1) at line 2. PPrune computes a PCS in $O(d|\mathcal{V}||PCS|)$ time (Theorem 4.1). Next, a partial CCS, $\mathcal{V}^*$, is constructed as follows: a random vector $\mathbf{V}$ from $\mathcal{V}'$ is selected at line 5. For $\mathbf{V}$, the algorithm tries to find a weight vector $\mathbf{w}$ for which $\mathbf{V}$ is better than the vectors in $\mathcal{V}^*$ (line 6), by solving the linear program in Algorithm 4.3. If there is such a $\mathbf{w}$, CPrune finds the best vector $\mathbf{V}'$ for $\mathbf{w}$ in $\mathcal{V}'$ and moves it to $\mathcal{V}^*$ (lines 10–12). If there is no weight for which $\mathbf{V}$ is better, it is C-dominated and thus removed from $\mathcal{V}'$ (line 8).

Theorem 4.2   *The computational complexity of CPrune as defined by Algorithm 4.2 is*

$$O(d|\mathcal{V}||PCS| + |PCS|P(d|CCS|)), \tag{4.5}$$

where $P(d|CCS|)$ is a polynomial in the size of the CCS and the number of objectives d, which is the runtime of the linear program that tests for C-domination (Algorithm 4).

To address (b), we can re-implement the algorithm for our simple example problem by substituting LCS for CPrune in Equation 4.3:

$$CCS = \text{CPrune}(\text{CPrune}(\mathcal{U}_l) \oplus \text{CPrune}(\mathcal{U}_r)).$$

To address (c), i.e., to confirm that the algorithm is correct, we provide a proof sketch for the simple MODP of Table 4.1. First, note that for each $\mathbf{w}$, CPrune retains at least one optimal value vector, and removes all vectors that are not optimal for any $\mathbf{w}$. Because linear scalarization distributes over addition, i.e., $\mathbf{w} \cdot (\mathbf{u}_l(a_l) + \mathbf{u}_r(a_r)) = \mathbf{w} \cdot \mathbf{u}_l(a_l) + \mathbf{w} \cdot \mathbf{u}_r(a_r)$, we know that no optimal values can be lost when performing CPrune on the separate sets $\mathcal{U}_l$ and $\mathcal{U}_r$, before taking the cross-sum. Furthermore, we know that because CPrune is applied again after taking the cross-sum, no excess value vectors can remain.

---

[1]Feng and Zilberstein [2004] proposed this method in a POMDP context. We explain the relation between POMDPs and multi-objective decision making in Section 5.2.1. Note that many algorithms for CCS pruning originate in other fields such as graphics or geometry, e.g., Graham [1972].

Algorithm 4.2 CPrune($\mathcal{V}$)

Input : A set of value vectors $\mathcal{V}$
Output : A CCS $\mathcal{V}^*$

1: $\mathcal{S} = \{\emptyset\}$
2: $\mathcal{V}' \leftarrow$ PPrune($\mathcal{V}$)
3: $\mathcal{V}^* \leftarrow \emptyset$
4: **while** $\mathcal{V}' \neq \emptyset$ **do**
5:      select random $\mathbf{V}$ from $\mathcal{V}'$
6:      $\mathbf{w} \leftarrow$ findWeight($\mathbf{V}, \mathcal{V}^*$)
7:      **if** $\mathbf{w}$=null **then**
8:          remove $\mathbf{V}$ from $\mathcal{V}'$
9:      **else**
10:          $\mathbf{V}' \leftarrow \text{argmax}_{\mathbf{V} \in \mathcal{V}'} \mathbf{w} \cdot \mathbf{V}$
11:          $\mathcal{V}' \leftarrow \mathcal{V}' \setminus \{\mathbf{V}'\}$
12:          $\mathcal{V}^* \leftarrow \mathcal{V}^* \cup \{\mathbf{V}'\}$
13:      **end if**
14: **end while**
15: **return** $\mathcal{V}^*$

Algorithm 4.3 findWeight($\mathbf{V}, \mathcal{V}$)

Input : A candidate value vector $\mathbf{V}$, and set of value vectors $\mathcal{V}$
Output : A weight vector $\mathbf{w}$ where $\mathbf{V}$ has a higher scalarized value than any vector in $\mathcal{V}$

$$\max_{x, \mathbf{w}} \quad x$$
$$\text{subject to} \quad \mathbf{w} \cdot (\mathbf{V} - \mathbf{V}') - x \geq 0, \forall \mathbf{V}' \in \mathcal{V}$$
$$\sum_{i=1}^{d} w_i = 1$$

if $(x > 0)$ {**return** $\mathbf{w}$} **else** { **return** null }

4.1.4   DESIGN CONSIDERATIONS

We have now defined our first inner loop algorithms, which find a PCS or a CCS for the problem of Table 4.1. However, these two algorithms are not the only possible ones. For example, consider the design choices we made for (b) for finding the CCS, i.e., applying CPrune at every possible location, and note that, when applying CPrune to $\mathcal{U}_l$ and $\mathcal{U}_r$, it removes one vector each. In other words, the local CCSs both consist of three vectors. Therefore, the cross-sum of these CCSs consists of 9 vectors, which is significantly less than the 16 we would get if we took the cross-sum before local pruning. However, in order to do the local pruning, we have to invest effort, which in this (small) problem amounts to as much work as computing the cross-sum. Therefore, we have to be careful when applying the inner loop approach; it is not always worth the effort to prune everywhere we can. Selecting how much to prune—by choosing either to not prune at all, apply only PPrune, or to apply CPrune, at each possible point in the algorithm—is a non-trivial and often problem-specific design choice.

Of course, in a small problem, such design choices are not critical, as long as the algorithm is correct. However, for complex problems, these design choices can have a profound impact on both the runtime and the memory usage of inner loop algorithms.

4.2   INNER LOOP PLANNING FOR MO-COGS

In cooperative multi-agent settings, agents must coordinate their behavior in order to optimize their common team payoff. Key to making coordination between agents efficient is exploiting the *loose couplings* common to such tasks: each agent's actions directly affect only a subset of the other agents. *Multi-objective coordination graphs* (MO-CoGs) (Section 2.2) express such independence in a graphical model for single-shot decisions.

In this section, we discuss inner loop planning methods for MO-CoGs that build on the single-objective *variable elimination* (VE) algorithm [Dechter, 1998, Guestrin et al., 2002, Kok and Vlassis, 2006, Rosenthal, 1977] for CoGs. First, the maximization operators of VE are replaced by cross-sums. Then, we show that by choosing the appropriate pruning operators, we can obtain algorithms that correctly compute either a PCS [Rollón, 2008, Rollón and Larrosa, 2006] or a CCS [Roijers et al., 2013b, 2015b], yielding a family of *multi-objective variable elimination* (MOVE) algorithms.

4.2.1   VARIABLE ELIMINATION

We first discuss single-objective variable elimination, on which several multi-objective algorithms build. VE exploits the loose couplings of a CoG (Definition 2.4) expressed by the local payoff functions to efficiently compute the optimal joint action **a** that maximizes the team payoff, $u(\mathbf{a})$.

First, in the *forward pass*, VE eliminates each of the agents in turn by computing the value of that agent's *best response* to every possible joint action of its neighbors. These best responses are used to construct a new local payoff function that encodes the values of the best responses. The new local payoff function then replaces the agent and the payoff functions in which it participated.

In the original algorithm, once all agents are eliminated, a *backward pass* assembles the optimal joint action using the constructed payoff functions. Here, we present a slight variant in which each payoff is 'tagged' with the action that generates it, obviating the need for a backward pass. While the two algorithms are equivalent, this variant is more amenable to the multi-objective (inner loop) extension we present in Section 4.2.3.

VE eliminates agents from the graph in a predetermined order. This order is typically determined heuristically (e.g., following Koller and Friedman [2009]), because finding the optimal elimination order is itself NP-hard [Arnborg, 1985, Dechter, 1998].

Algorithm 4.4 shows pseudocode for the elimination of a single agent i. First, VE determines the set, $\mathcal{U}_i$, of *local payoff functions* connected to i, and the set, n_i, of *neighboring agents* of i (lines 1–2).

Definition 4.3 The set, $\mathcal{U}_i$, of local payoff functions of i is the set of all local payoff functions that have agent i in scope.

Definition 4.4 The set, n_i, of neighboring agents of i is the set of all agents that are in scope of one or more of the local payoff functions in $\mathcal{U}_i$.

Then, VE constructs a new local payoff function, $u^{new}(\mathbf{a}_{n_i})$, by computing the value of agent i's best response to each possible joint action $\mathbf{a}_{n_i}$ of the agents in n_i (lines 3–12). To do so, it loops over all these joint actions $\mathcal{A}_{n_i}$ (line 4). For each $\mathbf{a}_{n_i}$, it loops over all the actions $\mathcal{A}_i$ available to agent i (line 6). For each $a_i \in \mathcal{A}_i$, it computes the local payoff when agent i responds to $\mathbf{a}_{n_i}$ with a_i (line 7). VE tags the total payoff with a_i, the action that generates it (line 8) in order to be able to retrieve the optimal joint action later. If there are already tags present, VE appends a_i to these tags. In this manner, the entire joint action is constructed incrementally. VE maintains the value of the best response by taking the maximum of these payoffs, and storing this maximal payoff in the new local payoff function, $u^{new}(\mathbf{a}_{n_i})$, on line 11. After the value for each $\mathbf{a}_{n_i}$ is computed, VE eliminates the agent and all payoff functions in $\mathcal{U}_i$ and replaces them with the newly constructed local payoff function (line 13).

Now, let us consider the example of Figure 2.1. The payoffs for this example were given in Table 2.1. The optimal payoff maximizes the sum of the two payoff functions:

$$\max_{\mathbf{a}} u(\mathbf{a}) = \max_{a_1, a_2, a_3} u^1(a_1, a_2) + u^2(a_2, a_3).$$

Figure 4.1 illustrates the application of VE to this problem, for the elimination order 3, 2, 1. VE eliminates agent 3 first, by pushing the maximization over a_3 inward such that goes only over the local payoff functions involving agent 3, in this case just u^2:

$$\max_{\mathbf{a}} u(\mathbf{a}) = \max_{a_1, a_2} \left(u^1(a_1, a_2) + \max_{a_3} u^2(a_2, a_3) \right).$$

Algorithm 4.4 `elimVE(`$\mathcal{U}, i$`)`

Input : A set, $\mathcal{U}$, of local payoff functions, and an agent i
Output : An updated set of local payoff functions from which i has been eliminated.

1: $\mathcal{U}_i \leftarrow$ set of local payoff functions involving i
2: $n_i \leftarrow$ set of neighboring agents of i
3: $u^{new} \leftarrow$ a new factor taking joint actions of n_i, $\mathbf{a}_{n_i}$, as input
4: **for all** $\mathbf{a}_{n_i} \in \mathcal{A}_{n_i}$ **do**
5: $\quad S \leftarrow \emptyset$ //set of action values for i
6: $\quad$ **for all** $a_i \in \mathcal{A}_i$ **do**
7: $\qquad v \leftarrow \displaystyle\sum_{u_j \in \mathcal{U}_i} u^j(\mathbf{a}_{n_i}, a_i)$
8: $\qquad$ tag v with a_i
9: $\qquad S \leftarrow S \cup \{v\}$
10: $\quad$ **end for**
11: $\quad u^{new}(\mathbf{a}_{n_i}) \leftarrow \max(S)$ //pick maximal value
12: **end for**
13: **return** $(\mathcal{U} \setminus \mathcal{U}_i) \cup \{u^{new}\}$

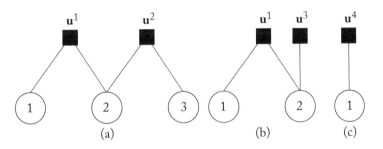

Figure 4.1: (a) A CoG with 3 agents and 2 local payoff functions, (b) after eliminating agent 3 by adding u^3, (c) after eliminating agent 2 by adding u^4.

VE solves the inner maximization and replaces it with a new local payoff function u^3 that depends only on agent 3's neighbors, thereby eliminating agent 3:

$$\max_{\mathbf{a}} u(\mathbf{a}) = \max_{a_1, a_2} \left(u^1(a_1, a_2) + u^3(a_2) \right),$$

which leads to the new factor graph depicted in Figure 2.1b. The values of $u^3(a_2)$ are $u^3(\dot{a}_2) =$ 2.5, using $\dot{a}_3$, and $u^3(\bar{a}_2) = 1$ using $\bar{a}_3$, as these are the optimal payoffs for the actions of agent 2, given the payoffs shown in Table 2.1.

Because we ultimately want the optimal joint action as well as the optimal payoff, VE needs to store the actions of agent 3 that correspond to the values in the new local payoff factor. In our adaption of VE, the algorithm tags each payoff of u^3 with the action of agent 3 that generates it. We can thus think of $u^3(a_2)$ as a tuple of value and tags, where value is a scalar payoff, and tags is a list of individual agent actions. We denote such a tuple with parentheses and a subscript: $u^3(\dot{a}_2) = (2.5)_{\dot{a}_3}$, and $u^3(\bar{a}_2) = (1)_{\bar{a}_3}$.

VE next eliminates agent 2, yielding the factor graph shown in Figure 2.1c:

$$\max_{\mathbf{a}} u(\mathbf{a}) = \max_{a_1}\left(\max_{a_2} u^1(a_1, a_2) + u^3(a_2)\right) = \max_{a_1} u^4(a_1).$$

VE appends the new tags for agent 2 to the existing tags for agent 3, yielding: $u^4(\dot{a}_1) = \max_{a_2} u^1(\dot{a}_1, a_2) + u^3(a_2) = (3.25)_{\dot{a}_2} + (2.5)_{\dot{a}_2 \dot{a}_3} = (5.75)_{\dot{a}_2 \dot{a}_3}$ and $u^4(\bar{a}_1) = (3.75)_{\bar{a}_2} + (1)_{\bar{a}_2 \bar{a}_3} = (4.75)_{\bar{a}_2 \bar{a}_3}$. Finally, maximizing over a_1 yields the optimal payoff — $(5.75)_{\dot{a}_1 \dot{a}_2 \dot{a}_3}$ — with the optimal action contained in the tags.

The computational complexity of VE is exponential in the *induced width*, w,

Theorem 4.5   *[Guestrin et al., 2002] The computational complexity of VE is $O(n|\mathcal{A}_{max}|^w)$ where $|\mathcal{A}_{max}|$ is the maximal number of actions for a* single *agent and w is the induced width, i.e., the maximal number of neighboring agents of an agent plus one (the agent itself), at the moment when it is eliminated.*

The induced width is limited by the number of agents. When the factor graph is fully connected, i.e., every agent shares a local payoff function with every other agent, w is equal to n. In practice however, w is typically much smaller than n.

The space complexity of VE is also exponential in the induced width:

Theorem 4.6   *[Dechter, 1998] The space complexity of VE is $O(n |\mathcal{A}_{max}|^w)$.*

This space complexity arises because, for every agent elimination, a new local payoff function is created with $O(|\mathcal{A}_{max}|^w)$ fields (possible input actions). Since it is impossible to tell *a priori* how many of these new local payoff functions exist at any given time during the execution of VE, this needs to be multiplied by the total number of new local payoff functions created during a VE execution, which is n.

VE is designed to minimize runtime. In fact, VE is proven to have the best runtime guarantees within a large class of algorithms [Rosenthal, 1977]. There are other methods that focus on memory efficiency instead [Mateescu and Dechter, 2005], either by sacrificing runtime, or by sacrificing optimality. For an overview of these methods, and how to create multi-objective methods from them, see [Roijers, 2016].

4.2.2   TRANSFORMING THE MO-COG

Using VE as a basis, we can build an inner loop method by replacing the summations with cross-sums and the maximizations with pruning. However, to apply the inner loop approach, we first need to be able to work with sets of value vectors rather than single vectors. Therefore, we first translate the MO-CoG to a set of *value set factors* (VSFs), $\mathcal{F}$, instead of the set of local payoff functions $\mathcal{U}$. Each VSF, $f^e \in \mathcal{F}$, is a function mapping local joint actions, $\mathbf{a}_e$ to sets of payoff vectors. The initial VSFs are constructed from the local payoff functions, $\mathbf{u}^e \in \mathcal{U}$, such that

$$f^e(\mathbf{a}_e) = \{\mathbf{u}^e(\mathbf{a}_e)\}, \tag{4.6}$$

i.e., each VSF maps a local joint action to the *singleton set* containing only that action's local payoff.

Using these VSFs, we can now define the set of all possible (team) payoff vectors, $\mathcal{V}$ in terms of $\mathcal{F}$ by performing cross-sums over all VSFs in $\mathcal{F}$ for each joint action $\mathbf{a}$:

$$\mathcal{V}(\mathcal{F}) = \bigcup_{\mathbf{a}} \bigoplus_{f^e \in \mathcal{F}} f^e(\mathbf{a}_e),$$

where $\bigoplus_{f^e \in \mathcal{F}}$ is the cross-sum (Definition 4.2) across all VSFs.

A coverage set can now be calculated by applying a the appropriate pruning operator, i.e., PPrune (Algorithm 4.1) for computing a PCS and CPrune (Algorithm 4.2) for computing a CCS. For example, in the case of a CCS:

$$CCS(\mathcal{V}(\mathcal{F})) = \text{CPrune}\,(\mathcal{V}(\mathcal{F})) = \text{CPrune}\left(\bigcup_{\mathbf{a}} \bigoplus_{f^e \in \mathcal{F}} f^e(\mathbf{a}_e) \right). \tag{4.7}$$

A naive, *non-graphical approach* to compute the CCS would simply compute the righthand side of Equation 4.7, i.e., it would compute $\mathcal{V}(\mathcal{F})$ explicitly by looping over all actions, and for each action looping over all local VSFs, and then pruning that set down to a CCS. This corresponds to *flattening* a MO-CoG to a list of all possible payoff vectors, and then pruning that list.

Because the non-graphical approach requires explicitly enumerating all possible joint actions and calculating the payoffs associated with each one, it is intractable for all but the smallest numbers of agents, as the number of joint actions grows exponentially in the number of agents. Therefore, we need a smarter way to compute CSs. We can achieve this by interleaving pruning and cross-sums, the same way that VE interleaves maximizations and summations.

4.2.3   MULTI-OBJECTIVE VARIABLE ELIMINATION

Similar to VE for CoGs, we can solve MO-CoGs as a series of local subproblems by eliminating agents and manipulating the set of VSFs $\mathcal{F}$ that describe the MO-CoG, an inner loop approach we call *multi-objective variable elimination* (MOVE). By changing where and how we choose to prune, we can generate variants of MOVE such as Rollón and Larrosa's algorithm for computing a PCS, which we refer to as PMOVE [Rollón and Larrosa, 2006], or the CMOVE algorithm for computing a CCS [Roijers et al., 2013b, 2015b].

The key idea behind MOVE is to compute LCSs when eliminating an agent instead of a single best response (as in VE). When computing an LCS, the algorithm prunes away as many vectors as possible, minimizing the number of payoff vectors that are calculated at the global level. Minimizing the number of payoff vectors that are calculated greatly reduces computation time.

Eliminating Agents

First, we describe the `elim` operator for eliminating agents from a set of VSF. This operator corresponds to the `elim` operator used by VE (Algorithm 4.4) for eliminating agents in single-objective CoGs. We first need to update our definition of neighboring local payoff functions (Definition 4.3), to neighboring VSFs.

Definition 4.7 The set of neighboring VSFs $\mathcal{F}_i$ of i is the set of all local payoff functions that have agent i in scope.

The neighboring agents n_i of an agent i are now the agents in the scope of a VSF in $\mathcal{F}_i$, except for i itself, corresponding to Definition 4.4. For each possible local joint action of n_i, we now compute an LCS that contains the payoffs of the undominated responses of agent i, as the best response values of i. In other words, it is the CS of the subproblem that arises when considering only $\mathcal{F}_i$ and fixing a specific local joint action $\mathbf{a}_{n_i}$.

To compute the LCS, we must consider all payoff vectors of the subproblem, $\mathcal{V}_i$, and prune the dominated ones. This can be achieved by taking the cross-sum of all the VSFs in the local subproblem, and then pruning.

Definition 4.8 If we fix all actions in $\mathbf{a}_{n_i}$, but not a_i, the set of all payoff vectors is: $\mathcal{V}_i(\mathcal{F}_i, \mathbf{a}_{n_i}) = \bigcup_{a_i} \bigoplus_{f^e \in \mathcal{F}_i} f^e(\mathbf{a}_e)$, where $\mathbf{a}_e$ is formed from a_i and the appropriate part of $\mathbf{a}_{n_i}$.

Using Definition 4.8, we can formally define the LCS as the CS of $\mathcal{V}_i$:

Definition 4.9 A local CS, an LCS, is the C-undominated subset of $\mathcal{V}_i(\mathcal{F}_i, \mathbf{a}_{n_i})$:

$$LCS_i(\mathcal{F}_i, \mathbf{a}_{n_i}) = CS(\mathcal{V}_i(\mathcal{F}_i, \mathbf{a}_{n_i})).$$

Because the LCS_i is the CS of a specified set of vectors, we can compute the LCS by standard pruning algorithms, such as PPrune (Algorithm 4.1) or CPrune (Algorithm 4.2). Using LCSs, we can create a new VSF, f^{new}, conditioned on the actions of the agents in n_i:

$$\forall \mathbf{a}_{n_i} \ f^{new}(\mathbf{a}_{n_i}) = LCS_i(\mathcal{F}_i, \mathbf{a}_{n_i}).$$

The `elim` operator replaces the VSFs in $\mathcal{F}_i$ in $\mathcal{F}$ by this new factor:

$$\text{elim}(\mathcal{F}, i) = (\mathcal{F} \setminus \mathcal{F}_i) \cup \{f^{new}(\mathbf{a}_{n_i})\}.$$

Eliminating an agent reduces the number of agents and VSFs in the graph, and forms the cornerstone operator of MOVE. Before defining MOVE, however, we first prove the correctness of `elim`, for both the PCS and the CCS cases. Particularly (as described in Section 4.1), we must prove that no necessary vectors are lost when applying `elim`.

We show, for the case of using MOVE to compute a CCS, that the maximal scalarized payoff, for any $\mathbf{w}$, cannot be lost as a result of `elim`, i.e., no necessary payoff vectors for creating a CCS are lost.

Theorem 4.10   *When* `elim` *employs local CCS pruning, it preserves the CCS, i.e.,*

$$\forall i \ \forall \mathcal{F} \ CCS(\mathcal{V}(\mathcal{F})) = CCS(\mathcal{V}(\texttt{elim}(\mathcal{F}, i))).$$

Proof. By definition, the CCS of a MO-CoG contains at least one payoff vector that maximizes the scalarized value for every $\mathbf{w}$. Therefore, for each vector $\mathbf{u}(\mathbf{a})$ that is optimal for at least one $\mathbf{w}$, there must be a vector that achieves the same scalarized value for that $\mathbf{w}$:

$$\forall \mathbf{w} \quad \left(\mathbf{a} = \operatorname*{argmax}_{\mathbf{a} \in \mathcal{A}} \mathbf{w} \cdot \mathbf{u}(\mathbf{a}) \right) \implies$$
$$\exists \mathbf{a}' \quad \mathbf{u}(\mathbf{a}') \in CCS(\mathcal{V}(\mathcal{F})) \ \wedge \ \mathbf{w} \cdot \mathbf{u}(\mathbf{a}) = \mathbf{w} \cdot \mathbf{u}(\mathbf{a}'). \tag{4.8}$$

If and only if this is not the case, necessary values are lost.

First, we observe that for all joint actions $\mathbf{a}$ for which there is a $\mathbf{w}$ at which the scalarized value of $\mathbf{a}$ is maximal, a vector-valued payoff $\mathbf{u}(\mathbf{a}')$ for which $\mathbf{w} \cdot \mathbf{u}(\mathbf{a}') = \mathbf{w} \cdot \mathbf{u}(\mathbf{a})$ is in the CCS (by definition). Second, we observe that the linear scalarization function distributes over the local payoff functions: $\mathbf{w} \cdot \mathbf{u}(\mathbf{a}) = \mathbf{w} \cdot \sum_e \mathbf{u}^e(\mathbf{a}_e) = \sum_e \mathbf{w} \cdot \mathbf{u}^e(\mathbf{a}_e)$. Thus, when eliminating agent i, we divide the set of VSFs into non-neighbors (nn), in which agent i does not participate, and neighbors (n_i) such that:

$$\mathbf{w} \cdot \mathbf{u}(\mathbf{a}) = \sum_{e \in nn} \mathbf{w} \cdot \mathbf{u}^e(\mathbf{a}_e) + \sum_{e \in ni} \mathbf{w} \cdot \mathbf{u}^e(\mathbf{a}_e).$$

Now, following (4.8), the CCS contains $\max_{\mathbf{a} \in \mathcal{A}} \mathbf{w} \cdot \mathbf{u}(\mathbf{a})$ for all $\mathbf{w}$. `elim` pushes this maximization in:

$$\max_{\mathbf{a} \in \mathcal{A}} \mathbf{w} \cdot \mathbf{u}(\mathbf{a}) = \max_{\mathbf{a}_{-i} \in \mathcal{A}_{-i}} \sum_{e \in nn} \mathbf{w} \cdot \mathbf{u}^e(\mathbf{a}_e) + \max_{a_i \in \mathcal{A}_i} \sum_{e \in ni} \mathbf{w} \cdot \mathbf{u}^e(\mathbf{a}_e).$$

`elim` replaces the agent-i factors by a term $f^{new}(\mathbf{a}_{n_i})$ that satisfies $\mathbf{w} \cdot f^{new}(\mathbf{a}_{n_i}) = \max_{a_i} \sum_{e \in n_i} \mathbf{w} \cdot \mathbf{u}^e(\mathbf{a}_e)$—for all $\mathbf{w}$—per definition, thus preserving the maximum scalarized value for all $\mathbf{w}$ and thereby preserving the CCS. $\square$

Instead of a *local CCS*, we could compute a *local PCS*.

Theorem 4.11   *When* `elim` *employs local PCS pruning, it preserves the PCS, i.e.,*

$$\forall i \ \forall \mathcal{F} \ PCS(\mathcal{V}(\mathcal{F})) = PCS(\mathcal{V}(\texttt{elim}(\mathcal{F}, i))).$$

Proof. First, note that for all joint actions **a**, the payoff vector is a sum of the local payoffs that form a subproblem for i, and those that are not in this subproblem:

$$\max_{\mathbf{a}\in\mathcal{A}} \mathbf{w}\cdot\mathbf{u}(\mathbf{a}) = \max_{\mathbf{a}_{-i}\in\mathcal{A}_{-i}} \sum_{e\in nn} \mathbf{w}\cdot\mathbf{u}^e(\mathbf{a}_e) + \max_{a_i\in\mathcal{A}_i} \sum_{e\in ni} \mathbf{w}\cdot\mathbf{u}^e(\mathbf{a}_e).$$

`elim` replaces the factors neighboring agent i with a term $f^{new}(\mathbf{a}_{n_i})$, conditioning on $\mathbf{a}_{n_i}$. Before this elimination happens, the solutions in $PCS(\mathcal{V}(\mathcal{F}))$ all consist of a component independent of i, i.e., the non-neighboring part, and a neighboring part. Each of these vectors has a specific value of $\mathbf{a}_{n_i}$. Because $f^{new}(\mathbf{a}_{n_i})$ conditions on $\mathbf{a}_{n_i}$, no necessary values of $\mathbf{a}_{n_i}$ can be lost. For a given $\mathbf{a}_{n_i}$, the non-neighboring part is independent of i, and we can write the total value as $\mathbf{v} + \mathbf{u_i}$, where $\mathbf{u_i} \in \mathcal{V}_i(\mathcal{F}_i, \mathbf{a}_{n_i})$. However, we know that, if there exists a $\mathbf{u_i}' \in \mathcal{V}_i(\mathcal{F}_i, \mathbf{a}_{n_i})$, for which $\mathbf{u_i}' \succ_P \mathbf{u_i}$, then $\mathbf{v} + \mathbf{u_i}$ cannot be undominated, as we can replace $\mathbf{u_i}$ by $\mathbf{u_i}'$, yielding $\mathbf{v} + \mathbf{u_i}' \succ_P \mathbf{v} + \mathbf{u_i}$. Therefore, we know that $f^{new}(\mathbf{a}_{n_i}) = PCS_i(\mathcal{F}_i, \mathbf{a}_{n_i})$ cannot delete vectors that are P-undominated, i.e., those necessary for a PCS of the MO-CoG.   □

The choice between a CCS or a PCS should be made on the basis of the taxonomy presented in the previous chapter. When doing so, note that, since LCCS $\subseteq$ LPCS $\subseteq \mathcal{V}_i$, `elim`, computing a CCS not only reduces the problem size with respect to $\mathcal{V}_i$, it can do so more than would be possible if we only considered P-dominance. Therefore, focusing on the CCS can greatly reduce the sizes of local subproblems. Since the solution of a local subproblem is the input for the next agent elimination, the size of subsequent local subproblems is also reduced, which can lead to considerable speedups. Therefore, we argue that the CCS should be preferred over the PCS whenever possible.

Multi-Objective Variable Elimination

Using `elim`, we now present the full MOVE algorithm. In our implementation, `elim` uses either `PPrune` or `CPrune` to compute the local CSs. Like VE, MOVE iteratively eliminates agents until none are left. However, our implementation of `elim` computes a CS and outputs the correct joint actions for each payoff vector in this CS, rather than a single joint action.

As previously mentioned, there are two MOVE algorithms in the literature. CMOVE [Roijers et al., 2013b, 2015b] is an extension to Rollón and Larrosa's [2006] Pareto-based extension of VE, which we refer to as PMOVE. The most important difference between CMOVE and PMOVE is that CMOVE computes a CCS, which typically leads to much smaller subproblems and thus much better computational efficiency. In addition, Roijers et al. [2015b] identify the three places in MOVE where pruning can take place, yielding a more flexible algorithm with different tradeoffs, which we also describe here. Finally, following Roijers et al. [2015b], we employ the tagging scheme instead of the backward pass, as in Section 4.2.1.

Algorithm 4.5 presents an abstract version of MOVE that leaves the pruning operators unspecified. Depending on preference, these pruning operators can be filled in with `PPrune` (Algorithm 4.1), or `CPrune` (Algorithm 4.2), or other algorithms for computing PCSs or CCSs.

Depending on which pruning operators are used in which points in the algorithm, Algorithm 4.5 correctly computes a correct CCS or PCS.

If more is known about the scalarization function, f (Definition 1.1), this could be incorporated in new pruning algorithms, which could be used here as well. For example, if it is known that objective 1 is worth as least twice as much as objective 2, inside a linear scalarization function, this could be incorporated into the pruning operators, without affecting the correctness of the algorithm. However, we limit this discussion to the general cases of the PCS and CCS.

MOVE first translates the problem into a set of *vector-set factors* (VSFs), $\mathcal{F}$ on line 1, according to (4.6). Next, MOVE iteratively eliminates agents using elim (line 2–5). The elimination order can be determined using techniques devised for single-objective VE (e.g., Koller and Friedman [2009]).

Algorithm 4.6 shows our implementation of elim, parameterized with two pruning operators, prune1 and prune2. These pruning operators correspond to two different pruning locations inside the operator that computes LCS_i: $\texttt{ComputeLCS}_{\texttt{i}}(\mathcal{F}_i, \mathbf{a}_{n_i}, \texttt{prune1}, \texttt{prune2})$.

$\texttt{ComputeLCS}_{\texttt{i}}$ is implemented as follows: first we define a new cross-sum-and-prune operator $A \hat{\oplus} B = \texttt{prune1}(A \oplus B)$. LCS_i applies this operator sequentially:

$$\texttt{ComputeLCS}_{\texttt{i}}(\mathcal{F}_i, \mathbf{a}_{n_i}, \texttt{prune1}, \texttt{prune2}) = \texttt{prune2}\left(\bigcup_{a_i} \hat{\bigoplus}_{f^e \in \mathcal{F}_i} f^e(\mathbf{a}_e)\right). \qquad (4.9)$$

prune1 is applied to each cross-sum of two sets, via the $\hat{\oplus}$ operator, leading to *incremental pruning* [Cassandra et al., 1997], i.e.,

$$A \hat{\oplus} B \hat{\oplus} C = \texttt{prune1}(A \oplus \texttt{prune1}(B \oplus C)).$$

prune2 is applied at a coarser level, after the union. MOVE applies elim iteratively until no agents remain, resulting in a CS. When there are no agents left, f^{new} on line 3 has no agents to condition on. In this case, we consider the actions of the neighbors to be a single empty action: $a_\emptyset$.

Pruning can also be applied at the very end, after all agents have been eliminated, which we call prune3. After all agents have been eliminated, the final factor is taken from the set of factors (line 6), and the single set, S contained in that factor is retrieved (line 7). Note that we use the empty action $a_\emptyset$ to denote the field in the final factor, as it has no agents in scope. Finally prune3 is called on S. In increasing level of coarseness, we thus have three pruning operators: incremental pruning (prune1), pruning after the union over actions of the eliminated agent (prune2), and pruning after all agents have been eliminated (prune3).

MOVE Variants

When either prune2 or prune3 (or both) compute CCSs, we obtain CMOVE [Roijers et al., 2015b], which correctly computes the CCS. When no pruning takes place for prune3 and

---

Algorithm 4.5 MOVE($\mathcal{U}$, prune1, prune2, prune3, q)

Input : A set of local payoff functions $\mathcal{U}$ and an elimination order q (a queue with all agents)
Output : A CS

1: $\mathcal{F} \leftarrow$ create one VSF for every local payoff function in $\mathcal{U}$
2: **while** $\mathbf{a}_{n_i} \in \mathcal{A}_{n_i}$ **do**
3:    $i \leftarrow$ q.dequeue()
4:    $\mathcal{F} \leftarrow$ elim($\mathcal{F}, i$, prune1, prune2)
5: **end while**
6: $f \leftarrow$ retrieve final factor from $\mathcal{F}$
7: $\mathcal{S} \leftarrow f(a_\emptyset)$
8: **return** prune3($\mathcal{S}$)

Algorithm 4.6 elim($\mathcal{F}, i$, prune1, prune2)

Input : A set of VSFs $\mathcal{F}$, and an agent i to eliminate
Output : A new set of VSFs from which i has been eliminated

1: $n_i \leftarrow$ the set of neighboring agents of i
2: $\mathcal{F}_i \leftarrow$ the subset of VSF that have i in scope
3: $f^{new}(\mathbf{a}_{n_i}) \leftarrow$ a new VSF
4: **for all** $\mathbf{a}_{n_i} \in \mathcal{A}_{n_i}$ **do**
5:    $f^{new}(\mathbf{a}_{n_i}) \leftarrow$ ComputeLCS$_i$($\mathcal{F}_i, \mathbf{a}_{n_i}$, prune1, prune2)
6: **end for**
7: $\mathcal{F} \leftarrow \mathcal{F} \setminus \mathcal{F}_i \cup \{f^{new}\}$
8: **return** $\mathcal{F}$

---

prune1 = prune2 = PPrune, we obtain PMOVE [Rollón and Larrosa, 2006], which correctly computes the PCS.

As mentioned earlier, the local coverage sets are input to the next subproblem in the agent-elimination sequence. A smaller local coverage set thus results in a smaller subsequent subproblem. Because it is always possible to have a CCS that is a subset of the smallest possible PCS, the subproblems in CMOVE are thus always smaller than the corresponding subproblems in PMOVE. An important insight is that this results in faster computation overall for CMOVE compared to PMOVE.

There are several ways to implement the pruning operators that lead to correct instantiations of CMOVE. Both PPrune (Algorithm 4.1) and CPrune (Algorithm 4.2) can be used, as long as either prune2 or prune3 is CPrune. Note that if prune2 computes the CCS, prune3 is not necessary.

For example, consider the following two variants: *basic CMOVE*, which does not use prune1 and prune3 and only prunes at prune2 using CPrune, and *incremental CMOVE*, which uses CPrune at both prune1 and prune2. The latter invests more effort in intermediate pruning, which can result in smaller cross-sums, and a resulting speedup. However, when only a few vectors can be pruned in these intermediate steps, this additional speedup may not occur, and the algorithm creates unnecessary overhead.

We can also compute a PCS first, using prune1 and prune2, and then compute the CCS with prune3, i.e., use PMOVE, and then prune the result of PMOVE down to a CCS. However, this is useful only for small problems for which a PCS is cheaper to compute than a CCS.

Analysis

We now analyze the correctness of MOVE. We assume that prune2 is implemented with CPrune (in the case of a CCS) or PPrune in the case of a PCS.

Theorem 4.12   *MOVE correctly computes a PCS or CCS.*

Proof. The proof works by induction on the number of agents. The base case is the original MO-CoG, where each $f^e(\mathbf{a}_e)$ from $\mathcal{F}$ is a singleton set. Then, since elim preserves the PCS or CCS (see Theorems 4.10 and 4.11), no necessary vectors are lost. Furthermore, no excess payoff vectors are retained. When the last agent is eliminated, only one factor remains. Since it is not conditioned on any agent actions and is the result of an *LCS* computation, it must contain one set: the appropriate *CS*. □

Example

Consider the example in Figure 2.1a, using the payoffs defined by Table 2.2, and apply CMOVE, using CPrune for prune2, and no pruning for prune1 and prune3, i.e., Basic CMOVE.

First, CMOVE creates the VSFs f^1 and f^2 from $\mathbf{u}^1$ and $\mathbf{u}^2$. To eliminate agent 3, it creates a new VSF $f^3(a_2)$ by computing the LCCSs for every a_2 and tagging each element of each set with the action of agent 3 that generates it. For $\dot{a}_2$, CMOVE first generates the set $\{(3,1)_{\dot{a}_3}, (1,3)_{\bar{a}_3}\}$. Since both of these vectors are optimal for some $\mathbf{w}$, neither is removed by pruning and thus $f^3(\dot{a}_2) = \{(3,1)_{\dot{a}_3}, (1,3)_{\bar{a}_3}\}$. For $\bar{a}_2$, CMOVE first generates $\{(0,0)_{\dot{a}_3}, (1,1)_{\bar{a}_3}\}$. CPrune determines that $(0,0)_{\dot{a}_3}$ is dominated and consequently removes it, yielding $f^3(\bar{a}_2) = \{(1,1)_{\bar{a}_3}\}$. CMOVE then adds f^3 to the graph and removes f^2 and agent 3, yielding the factor graph shown in Figure 2.1b.

CMOVE then eliminates agent 2 by combining f^1 and f^3 to create f^4. For $f^4(\dot{a}_1)$, CMOVE must calculate the LCCS of:

$$\left(f^1(\dot{a}_1, \dot{a}_2) \oplus f^3(\dot{a}_2) \right) \cup \left(f^1(\dot{a}_1, \bar{a}_2) \oplus f^3(\bar{a}_2) \right).$$

The first cross sum yields $\{(7,2)_{\dot{a}_2\dot{a}_3}, (5,4)_{\dot{a}_2\bar{a}_3}\}$ and the second yields $\{(1,1)_{\bar{a}_2\bar{a}_3}\}$. Pruning their union yields $f^4(\dot{a}_1) = \{(7,2)_{\dot{a}_2\dot{a}_3}, (5,4)_{\dot{a}_2\bar{a}_3}\}$. Similarly, for $\bar{a}_1$ taking the union yields $\{(4,3)_{\dot{a}_2\dot{a}_3}, (2,5)_{\dot{a}_2\bar{a}_3}, (4,7)_{\bar{a}_2\bar{a}_3}\}$, of which the LCCS is $f^4(\bar{a}_1) = \{(4,7)_{\bar{a}_2\bar{a}_3}\}$. Adding f^4 results in the graph in Figure 2.1c.

Finally, CMOVE eliminates agent 1. Since there are no neighboring agents left, $\mathcal{A}_i$ contains only the empty action. CMOVE takes the union of $f^4(\dot{a}_1)$ and $f^4(\bar{a}_1)$. Since $(7,2)_{\{\dot{a}_1\dot{a}_2\dot{a}_3\}}$ and $(4,7)_{\{\bar{a}_1\bar{a}_2\bar{a}_3\}}$ dominate $(5,4)_{\{\dot{a}_1\dot{a}_2\bar{a}_3\}}$, the latter is pruned, leaving $CCS = \{(7,2)_{\{\dot{a}_1\dot{a}_2\dot{a}_3\}}, (4,7)_{\{\bar{a}_1\bar{a}_2\bar{a}_3\}}\}$.

4.2.4   COMPARING PMOVE AND CMOVE

We now compare PMOVE and CMOVE. We already mentioned the intuition that the smaller the subproblems, the more efficient the algorithm. Here, we first provide a theoretical comparison, on the basis of a runtime analysis, and then an experimental comparison using an illustrative problem.

Theoretical Comparison
We first analyze the complexity of MOVE.

Theorem 4.13   *The computational complexity of MOVE is*

$$O\left(n \, |\mathcal{A}_{max}|^{w_a} \left(w_f \, R_1 + R_2 \right) + R_3 \right), \tag{4.10}$$

where w_a is the induced agent width, i.e., the maximum number of neighboring agents (connected via factors) of an agent when eliminated, w_f is the induced factor width, i.e., the maximum number of neighboring factors of an agent when eliminated, and R_1, R_2, and R_3 are the cost of applying the prune1, prune2, *and* prune3 *operators.*

Proof. MOVE eliminates n agents and for each one computes an LCS for each joint action of the eliminated agent's neighbors, in a field in a new VSF. MOVE computes $O(|\mathcal{A}_{max}|^{w_a})$ fields per iteration, calling prune1 (see (4.9)) for each adjacent factor, and prune2 once after taking the union over actions of the eliminated agent. prune3 is called exactly once, after eliminating all agents (line 8 of Algorithm 4.5). □

R_1, R_2, and R_3—the runtimes of the pruning operators—depend on the size of the local subproblems. Specifically, the runtime of PPrune is $O(d|\mathcal{V}||PCS|)$ (Theorem 4.1) and the runtime of CPrune is $O(d|\mathcal{V}||PCS| + |PCS|P(d|CCS|))$ (Theorem 4.2), where $|\mathcal{V}|$ is the number of vectors inputted to the pruning algorithm. The sizes of these input sets depend on how much pruning has been done in earlier iterations and at finer levels. Specifically, the input set of prune2 is the union of what is returned by a series of applications of prune1, while prune3 uses the output of the last application of prune2. We therefore need to balance the effort of the lower-level pruning with that of the higher-level pruning, which occurs less often but is dependent on the output of the lower level. The bigger the local CSs, the more can be gained from lower-level pruning.

MOVE is exponential only in w_a, and not in the number of agents as in the non-graphical approach. However, by describing the computational complexity in a way that makes explicit the dependence on the effectiveness of pruning, i.e., in terms of R_1, R_2, and R_3, we can see that MOVE is typically much faster than a non-graphical approach; in the worst case, when no pruning is possible, the complexity bounds are not better; but we can see the improvement clearly when pruning *is* effective.

When comparing PMOVE and CMOVE, the effects of pruning are highly relevant. While the runtime of PPrune, i.e., $O(d|\mathcal{V}||PCS|)$, might seem smaller than that of CPrune, i.e., $O(d|\mathcal{V}||PCS| + |PCS|P(d|CCS|))$, the effect of pruning is on the size of the input to these algorithms. For PMOVE, the input size, $|\mathcal{V}|$, for the pruning operator is the result of previous local PCS computations, while for CMOVE, it is the result of local CCS computations. Because CCSs are (typically much smaller) subsets of PCSs, CMOVE is faster than PMOVE when the coverage set size is sufficiently large. Because the coverage set size grows with the size of the MO-CoG, CMOVE is thus faster than PMOVE for larger problems.

Besides the computational complexity, another important aspect of the behavior of MO-CoG algorithms is the space complexity. In fact, memory is often the bottleneck for solving MO-CoGs.

Theorem 4.14   *The space complexity of CMOVE is*

$$O(\ d\ n\ |\mathcal{A}_{max}|^{w_a}\ |LCS_{max}| +\ d\ \rho\ |\mathcal{A}_{max}|^{|e_{max}|} + |LCS_{max}|^{|\mathcal{F}_{max}|}\ ),$$

where $|LCS_{max}|$ is maximum size of a local CCS, ρ is the original number of VSFs, $|e_{max}|$ is the maximum scope size of the original VSFs, and the $|\mathcal{F}_{max}|$ is the maximal number of VSFs that get replaced by a new VSF.

Proof. MOVE computes a local CS—using PPrune or CPrune—for each new VSF for each joint action of the eliminated agent's neighbors. There are maximally w_a neighbors. There are maximally n new VSFs. Each payoff vector stores d real numbers.

The size of the input of the pruning operator depends on the number of VSFs, $|\mathcal{F}_{max}|$, that get replaced by a new VSF, and the size of the local CSs in the VSFs that are replaced, i.e., the size of the cross-sum in (4.9). Note that, in the case of incremental pruning (i.e., instantiating `prune1` with a pruning operator), the size of this cross-sum is limited even further, by pruning after each cross-sum of two local CSs.

There are ρ VSFs created during the initialization of MOVE. All of these VSFs have exactly one payoff vector containing d real numbers, per joint action of the agents in scope. There are maximally $|\mathcal{A}_{max}|^{|e_{max}|}$ such joint actions. $\square$

Note that Theorem 4.14 is a loose bound because not all VSFs, original or new, exist at the same time. However, it is not possible to predict *a priori* how many of these VSFs exist at the same time, resulting in a space complexity bound on the basis of all VSFs that exist at some point during the execution of MOVE.

For PMOVE and CMOVE, the space complexity is equivalent to Theorem 4.14. However, for PMOVE, the size of $|LPS_{max}|$ is that of a local PCS, $|LPCS_{max}|$, while for CMOVE it is the size of an LCCS, $|LCCS_{max}|$. Because an LCCS is a subset of the corresponding LPCS,[2] CMOVE is thus strictly more memory efficient than PMOVE.

Experimental Comparison

To illustrate the difference in computational efficiency between PMOVE and CMOVE, we present results from an experimental evaluation on the Mining Day problem [Roijers et al., 2015b].

Mining Day is inspired by the example problem described in Chapter 1 and illustrated in Figure 1.2: a mining company mines gold and silver (objectives) from a set of mines (local payoff functions) located in the mountains. The mine workers live in villages at the foot of the mountains. The company has one van in each village (agents) for transporting workers and must determine every morning to which mine each van should go (actions). However, vans can only travel to nearby mines (graph connectivity). Workers are more efficient if there are more workers at the mine: there is a 3% efficiency bonus per worker such that the amount of each resource mined per worker is $x \cdot 1.03^w$, where x is the base rate per worker and w is the number of workers at the mine. The base rate of gold and silver are properties of a mine. Since the company aims to maximize revenue, the best strategy depends on the fluctuating prices of gold and silver. To maximize revenue, the mining company wants to use the latest possible price information, and not lose time recomputing the optimal strategy with every price change. Therefore, we must calculate a CCS.

To generate a Mining Day instance with v villages (agents), we randomly assign 2–5 workers to each village and connect it to 2–4 mines. Each village is only connected to mines with a greater or equal index, i.e., if village i is connected to m mines, it is connected to mines i to $i + m - 1$. The last village is connected to 4 mines and thus the number of mines is $v + 3$. The

---

[2]Our implementation of `CPrune` (Algorithm 4.2) uses `PPrune` as a pre-processing step (on line 2).

base rates per worker for each resource at each mine are drawn uniformly and independently from the interval [0, 10].

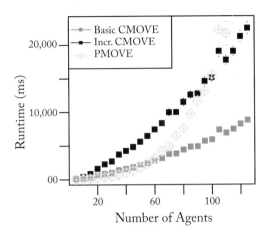
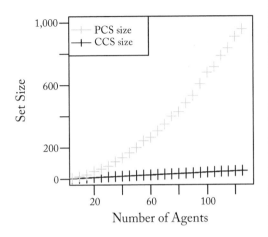

Figure 4.2: Runtimes (ms) for basic and incremental CMOVE, and PMOVE, in log-scale with the standard deviation of mean (error bars) (left) and the corresponding number of vectors in the PCS and CCS (right), for increasing numbers of agents.

In order to compare the runtimes of basic and incremental CMOVE against PMOVE, we generate Mining Day instances with varying numbers of agents. The runtime results are shown in Figure 4.2 (left). Both CMOVE and PMOVE are able to tackle problems with over 100 agents. However, the runtime of PMOVE grows much more quickly than that of CMOVE. In this two-objective setting, basic CMOVE is better than incremental CMOVE. However, not that this is problem dependent: when the CCSs are larger, it may well be worth it to invest more time in pruning in intermediate computations. Also note that more pruning in intermediate steps generates smaller intermediate sets of value vectors, and is thus more memory-efficient as well.

Basic CMOVE and PMOVE both have runtimes of around 2.8s at 60 agents, but at 100 agents, basic CMOVE runs in about 5.9s and PMOVE in 21s. Even though incremental CMOVE is worse than basic CMOVE, its runtime still grows much more slowly than that of PMOVE, and it beats PMOVE when there are many agents.

The difference between PMOVE and CMOVE results from the relationship between the number of agents and the sizes of the CCS, which grows linearly, and the PCS, which grows polynomially, as shown in Figure 4.2 (right). The induced width remains around 4 regardless of the number of agents.

These results demonstrate that, as the CCS grows more slowly than the PCS with the number of agents, CMOVE can solve MO-CoGs more efficiently than PMOVE as the number of agents increases. This is generally true across inner loop methods: the PCS-version of the inner loop method is more efficient for a small number of agents, while the CCS version is more efficient for larger numbers of agents. Furthermore, CCS inner loop methods are also more memory-efficient than corresponding PCS methods. For an extensive comparison in terms of both runtime and memory usage, see [Roijers et al., 2015b].

4.3   INNER LOOP PLANNING FOR MOMDPS

In the previous section, we considered problems in which a single joint action is performed in order to obtain a reward. However, many decision problems consist of a sequence of decisions. This sequence of decisions typically takes place in an environment that is affected by these decisions. Therefore, agents do not only have to consider their immediate reward, but also the reward they will be able obtain later, by changing the state of the environment to a more favorable one.

In this section, we discuss inner loop planning methods for multi-objective Markov decision processes (Definiton 2.8) that build on the single-objective *value iteration* (VI) algorithm [Bellman, 1957a, Sutton and Barto, 1998, Wiering and Van Otterlo, 2012] for MDPs (Definition 2.6).

The inner loop approach follows the same procedure used to convert VE to MOVE in the previous section: we first restate the problem in terms of sets of value vectors, replace the summations by cross-sums, and then identify places to prune the intermediate sets of value vectors.

4.3.1   VALUE ITERATION

Before we discuss the methods for planning in MOMDPs, we first discuss the single-objective algorithm upon which they build. Value iteration takes the Bellman optimality equation (see (2.5)) and turns it into a *Bellman backup*:

$$V_{k+1}(s) \leftarrow \max_a \sum_{s'} T(s, a, s')[R(s, a, s') + \gamma V_k(s')], \qquad (4.11)$$

where k indexes the current iteration of VI. At each iteration, all states are backed up once. When $k \to \infty$, starting from any initial value $V_0(s)$, VI will converge to the optimal value function $V^*(s)$.

For convenience, we can split Equation (4.11) into two parts:

$$V_{k+1}(s) \leftarrow \max_a Q_{k+1}(s, a), \qquad (4.12)$$

$$Q_{k+1}(s, a) \leftarrow \sum_{s'} T(s, a, s')\Big[R(s, a, s') + \gamma V_k(s')\Big]. \qquad (4.13)$$

Using $Q_{k+1}(s, a)$, we can not only compute the next value $V_{k+1}(s)$ but also derive the greedy policy:

$$\pi_{k+1}(s) \leftarrow \operatorname*{argmax}_a Q_{k+1}(s, a). \qquad (4.14)$$

Note that in an MDP, there is always a deterministic stationary optimal policy, due to Theorem 2.7. Therefore, after VI has converged, Equation (4.14) yields an optimal policy.

4.3.2 MULTI-OBJECTIVE VALUE ITERATION

To create an inner loop planning method from VI, we first recast the problem such that the algorithm works on sets of value vectors instead of single scalar values. For *multi-objective VI* (MOVI), it is sufficient to set an initial set of value vectors. For example for a 2-objective MOMDP, we could use the singleton set with the zero-vector:

$$V_0(s) = \{(0, 0)\}.$$

Then, we update the backup operator to handle sets of value vectors, i.e., to generate all value vectors for each (s, a) pair and then prune away the dominated ones. First, we adapt the update rule for Q-values from (4.13):

$$\mathbf{Q}_{k+1}(s, a) \leftarrow \bigoplus_{s'} T(s, a, s') \left[\mathbf{R}(s, a, s') + \gamma \mathbf{V}_k(s') \right]$$

where $\mathbf{u} + V = \{\mathbf{u} + \mathbf{v} : \mathbf{v} \in V\}$, and $U \oplus V = \{\mathbf{u} + \mathbf{v} : \mathbf{u} \in U \wedge \mathbf{v} \in V\}$. Then, we prune value vectors, using a CS computation, when calculating the new set of value vectors (updating (4.12)):

$$\mathbf{V}_{k+1}(s) \leftarrow CS \left(\bigcup_a \mathbf{Q}_{k+1}(s, a) \right).$$

The CS computation can be implemented with standard pruning operators. When the CS computation is implemented using PPrune (Algorithm 4.1), we get the Pareto-based multi-objective value iteration (PMOVI) of White [1982]. When we use CPrune instead, we obtain *convex hull value iteration* (CHVI) [Barrett and Narayanan, 2008]. Both algorithms converge to the appropriate coverage set.

While the PCS contains values for deterministic non-stationary policies, it is possible to distill a CCS of deterministic stationary policies for the CCS. That is, for the CCS, we can input a linear scalarization weight vector $\mathbf{w}$, and retrieve an optimal deterministic stationary policy for each weight:

$$\pi_{k+1}(s, \mathbf{w}) \leftarrow \operatorname*{argmax}_a \max_{\mathbf{q} \in Q_{k+1}(s,a)} \mathbf{w} \cdot \mathbf{q}. \tag{4.15}$$

However, value vectors in the PCS may depend on taking different actions in the same state at different time-steps. Therefore, retrieving a policy requires an on-the-fly *policy-tracking procedure* that recalculates the values expected at the next timestep after executing each action [Van Moffaert, 2016].

4.3.3 PARETO VS. CONVEX VALUE ITERATION

We illustrate the multi-objective version of value iteration with a simple 2-objective MOMDP with:

- only one state: s;

- two actions: a_1 and a_2;

- deterministic transitions, i.e., $T(s, a_1, s) = T(s, a_2, s) = 1$;

- deterministic rewards:

    - $\mathbf{R}(s, a_1, s) \rightarrow (2, 0)$ and
    - $\mathbf{R}(s, a_2, s) \rightarrow (0, 2)$;

- an initial state distribution $\mu_0(s) = 1$; and

- a discount factor $\gamma = 0.5$.

We initialize the value function with a singleton set containing the zero-vector:

$$\mathbf{V}_0(s) = \{(0, 0)\}.$$

We now illustrate what happens when we run multi-objective value iteration, using both PPrune and CPrune. We use the value-space perspective (Figure 3.1 (left)) to illustrate the former, and the weight-space perspective (Figure 3.1 (right)) to illustrate the latter.

In the first iteration, we bootstrap off the only possible vector for $\mathbf{V}_0(s)$, i.e., $(0, 0)$, to compute the $\mathbf{Q}$-value sets: $\mathbf{Q}_1(s, a_1) = \{(2, 0)\}$ and $\mathbf{Q}_1(s, a_2) = \{(0, 2)\}$. Then, we compute the new set of value vectors, $\mathbf{V}_1(s)$, by either PPrune or CPrune. In both cases, neither vector can be pruned:

$$\mathbf{V}_1(s) = \text{CPrune}\left(\bigcup_a \mathbf{Q}_1(s, a)\right) = \text{PPrune}\left(\bigcup_a \mathbf{Q}_1(s, a)\right) = \{(2, 0), (0, 2)\}.$$

The vectors after the first iteration are displayed in Figure 4.3.

In the second iteration, the algorithms for computing the PCS and the CCS start to differ. The $\mathbf{Q}$ sets are still the same: $\mathbf{Q}_2(s, a_1) = \{(3, 0), (2, 1)\}$ and $\mathbf{Q}_2(s, a_2) = \{(1, 2), (0, 3)\}$, but after taking the union of the $\mathbf{Q}$ sets,

$$\mathbf{V}_{p2}(s) = \text{PPrune}(\{(3, 0), (2, 1), (1, 2), (0, 3)\}) = \{(3, 0), (2, 1), (1, 2), (0, 3)\},$$

while

$$\mathbf{V}_{c2}(s) = \text{CPrune}(\{(3, 0), (2, 1), (1, 2), (0, 3)\}) = \{(3, 0), (0, 3)\},$$

as the vectors $(2, 1)$ and $(1, 2)$ are not P-dominated (Figure 4.4 (left)) but are not necessary to construct a CCS, as they are optimal for only one linear scalarization weight $\mathbf{w}$, where $(3, 0)$ and $(0, 3)$ are optimal as well.

In the third iteration, Pareto-based VI starts with a larger set of value vectors than CHVI. After computing the $\mathbf{Q}$ sets, there are eight value vectors, none of which can be pruned by PPrune

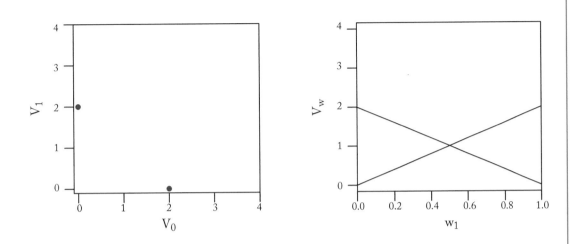

Figure 4.3: (Left) The result after the first iteration of Pareto VI (and CHVI) in value space: there are two vectors that are in both the intermediate PCS and the intermediate CCS. The red value vectors are in both the CCS and the PCS. (Right) The result of CHVI after the first iteration in weight space.

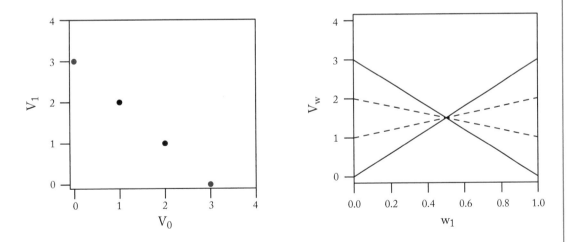

Figure 4.4: (Left) The result after the second iteration of Pareto VI (and CHVI) in value space: there are two vectors that are in both the intermediate PCS and the intermediate CCS indicated in red, and two vectors shown in black that are in the PCS but not in the CCS. (Right) The result of CHVI after the second iteration in weight space. The dotted lines indicate the vectors pruned by CPrune.

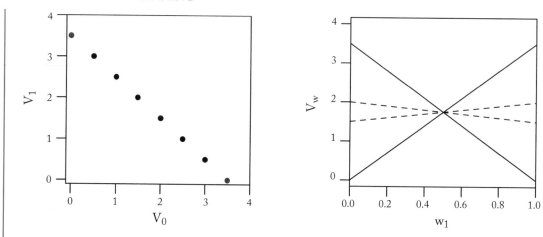

Figure 4.5: (Left) The result after the third iteration of Pareto VI (and CHVI) in value space: there are two vectors that are in both the intermediate PCS and the intermediate CCS indicated in red, and six vectors shown in black that are in the PCS but not in the CCS. (Right) The result of CHVI after the third iteration in weight space. The dotted lines indicate the vectors pruned by CPrune.

(Figure 4.5 (left)). CHVI on the other hand, generates four value vectors, two of which can be pruned, as indicated in Figure 4.5 (right).

If we continue, Pareto-based VI will double the number of vectors at each iteration, *ad infinitum*, because the PCS of deterministic non-stationary policies consists of infinitely many policies. CHVI on the other hand, will generate only four vectors and keep only two at each iteration, as the CCS requires only two deterministic stationary policies to cover all possible policies (including stochastic and non-stationary ones). The value vectors corresponding to these two policies are exactly those to which CHVI converges.

CHAPTER 5

Outer Loop Planning

In this chapter, we discuss the outer loop approach to solving multi-objective decision problems. In contrast to the inner loop approach—in which a single-objective algorithm for an SODP is adapted to apply to the corresponding MODP, by changing the summation and maximization operators into cross-sum and pruning operators—the outer loop approach leaves the single-objective algorithm intact. Instead, an MODP is solved as a series of scalarized (i.e., single-objective) problems, and single-objective algorithms are used as subroutines.

Outer loop approaches work by exploiting mathematical properties of the CCS. Consequently, they are primarily applicable only to CCS planning problems, which we focus on in this chapter. However, in Section 5.8, we briefly discuss their diminished applicability to PCS planning.

In Section 5.2, we begin to develop the outer loop approach by identifying the mathematical properties of the CCS and the linear scalarization function that can be used to identify the CCS. We define the *scalarized value function* given a CCS under linear scalarization, and observe that this function is *piecewise linear and convex (PWLC)*. This property can be exploited by outer loop methods to identify the CCS. We also explain the connection with the *partially observable Markov decision process*, a related problem in which the optimal value function is also PWLC (in Section 5.2.1).

In Section 5.3, we discuss a generic outer loop algorithm called *optimistic linear support* (OLS) [Roijers, 2016, Roijers et al., 2014b, 2015b], that exploits the PWLC property of the scalarized value function to identify the CCS correctly and in finite time. OLS takes a single-objective method as a subroutine, which it calls a finite number of times in order to solve an MODP. In Section 5.4, we analyze OLS theoretically, and show that it has many advantages over inner loop algorithms. Firstly, OLS comes with strong guarantees with respect to time and space complexity. Secondly, OLS-based algorithms can be much faster for small and medium numbers of objectives than corresponding inner loop algorithms. And finally, in OLS, the single-objective subroutines can be used out of the box, making any improvement for single-objective methods an improvement for multi-objective methods.

5.1 OUTER LOOP APPROACH

In an *outer loop approach*, an (approximate) coverage set is built incrementally, by solving scalarized instances. In order to solve these scalarized instances, a subroutine appropriate for the single-objective version of the MODP at hand is required. For example, when we want to solve an

MOMDP, an MDP solver is required. The solutions produced for scalarized instances are kept in a partial coverage set.

Definition 5.1   A *partial CS*, $\mathcal{S}$, is a subset of a CS, which is in turn a subset of all possible value vectors ($\mathcal{V} = \{\mathbf{V}^\pi : \pi \in \Pi\}$), i.e., $\mathcal{S} \subseteq CS \subseteq \mathcal{V}$.

The basic structure of an outer loop method is given in Algorithm 5.7. It starts from an empty set (line 1), as a partial CS. In each iteration, a scalarized instance (i.e., an SODP resulting from scalarizing the MODP with a given f and a $\mathbf{w}$) is selected (line 3), and solved using a single-objective subroutine (line 4). For each solution, i.e., an optimal policy $\pi_\mathbf{w}^*$, of a scalarized instance, the multi-objective value vector $\mathbf{V}_\mathbf{w}$ is retrieved. If $\mathbf{V}_\mathbf{w}$ improves upon the partial CS, $\mathcal{S}$—by improving the scalarized value for *some* (allowed) f and $\mathbf{w}$—it is added to $\mathcal{S}$ (lines 5–8). This process continues until some stop criterion is reached, e.g., when it can be proven that $\mathcal{S}$ is a CS or time runs out.

Algorithm 5.7 OuterLoopMethod(m, SolveSO)

Input : An MODP, m, and a corresponding single-objective solver SolveSO.
Output : A (partial) CS

1: $\mathcal{S} \leftarrow \emptyset$ //a partial CS
2: **while** stop criterion not reached **do**
3:      $m_\mathbf{w} \leftarrow$ select a scalarization weight $\mathbf{w}$ and scalarize m
4:      $\pi_\mathbf{w}^* \leftarrow$ SolveSO($m_\mathbf{w}$)
5:      $\mathbf{V}_\mathbf{w} \leftarrow$ retrieve/compute the multi-objective value of $\pi_\mathbf{w}^*$
6:      **if** $\mathbf{V}_\mathbf{w}$ improves upon S **then**
7:          $\mathcal{S} \leftarrow \mathcal{S} \cup \{\mathbf{V}_\mathbf{w}\}$ //add $\mathbf{V}_\mathbf{w}$ (and associated $\pi_\mathbf{w}^*$) to $\mathcal{S}$.
8:      **end if**
9: **end while**
10: **return** S

This outer loop approach is typically used for CCS planning, because the CCS is the optimal solution set given a linear scalarization function, and linear scalarization can be done without destroying the additivity of the returns in an MOMDP, or the additive sum over local payoff functions in a MO-CoG. For nonlinear f, it would be highly non-trivial to scalarize these MODPs to a suitable SODP, preempting effective use of the outer loop approach.

The three main design choices when creating an outer loop method are:

(a) how to select scalarized instances to solve,

(b) which single-objective subroutine to use, and

(c) whether to compute the value vectors separately (by policy evaluation), or to adapt the single-objective solver such that it returns both $\pi_\mathbf{w}^*$ and $\mathbf{V}_\mathbf{w}$.

An example of an outer loop approach is *random sampling* (RS). In RS, scalarized instances are selected randomly, by sampling one allowed $\mathbf{w}$ (and f) at each iteration and scalarizing the problem. The scalarized problem $m_\mathbf{w}$ is then solved using, e.g., an out-of-the-box single-objective solver that produces an optimal policy for $m_\mathbf{w}$, and the value vector, $\mathbf{V_w}$ retrieved by a standard policy evaluation algorithm. At each iteration, RS checks whether $\mathbf{V_w}$ is already in $\mathcal{S}$ or not—note that $\mathbf{V_w}$ cannot be dominated if the single-objective solver is optimal since by definition it is optimal for at least one scalarized instance—and if not, added to $\mathcal{S}$. Typically, this is done for a limited number of iterations, or until a fixed number I_{stop} of iterations has not produced a new value vector for $\mathcal{S}$.

RS has two important advantages. Firstly, it quickly produces *some* result, and improves upon it iteratively, i.e., it is an *anytime* algorithm. This is a major advantage over many inner loop methods, which typically need to run until completion to produce any useful results. Secondly, RS is easy to implement, as long as a suitable single-objective solver and a policy evaluation method are available.

However, RS also has a major downside: it is optimal only in the limit. Because it selects weights randomly, there is no guarantee within finite time that there is no f and $\mathbf{w}$ left for which an improvement still exists.

Instead of sampling randomly, *weighted-sum* methods [Kim and de Weck, 2005, Van Moffaert et al., 2014] sample on the basis of, e.g., relative distance to the value vectors $\mathbf{V}$ or the hypervolume metric. However, these methods still do not provide runtime or optimality guarantees.

In Section 5.3, we present a recent algorithm which retains the advantages of RS and weighted-sum methods, but is provably optimal when the single-objective subroutines used are optimal, and produces bounded approximations when the subroutines produce upper and lower bounds, and runs in finite time. It does so by exploiting mathematical properties of the CCS for MODPs and the function we obtain when we view the optimal linearly scalarized value given a CCS as a function of the scalarization weight $\mathbf{w}$.

5.2   SCALARIZED VALUE FUNCTIONS

To create provably optimal outer loop CCS planning methods that run in finite time and make smart choices about which scalarized instances of the MODP to solve, we first need to define the *scalarized value function* with respect to the linear scalarization function (Definition 3.1). Note that focusing on linear scalarization is sufficient to discover the CCS, even though the CCS can also be used in the context of nonlinear scalarization.

If we view the linearly scalarized value $\mathbf{V_w}$ of a value vector $\mathbf{V}$ as a function of $\mathbf{w}$, it becomes a hyperplane above the weight simplex. For example, consider the value vectors of the deterministic policies for the simple MO-CoG example of Section 4.1.1 (Table 4.1) shown in Figure 5.1. At left is the value space perspective, i.e., each policy is a point with the value in objective 0 on the x-axis, and objective 1 on the y-axis. At right is the weight space perspective, i.e., the scalarized value as a function of the weights, in this case just w_1. Because there are two objectives, and the scalarization function is a convex combination, $w_0 = 1 - w_1$. When plotting the scalarized value as a function of w_1, each value vector becomes a line. The policies in the CCS are shown in purple.

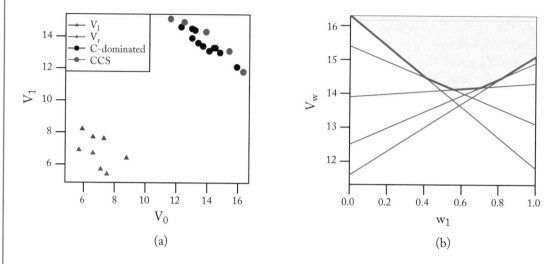

(a)   (b)

Figure 5.1: (a) All values and local rewards for the problem of Table 4.1. (b) The optimal scalarized value function, $V_{CCS}^*(\mathbf{w}) = \max_{\mathbf{V} \in CCS} \mathbf{w} \cdot \mathbf{V}$.

The maximal scalarized value function over a set of vectors, $\mathcal{S}$, takes the maximum over all the vectors in that set.

Definition 5.2   A scalarized value function over a partial CCS, $\mathcal{S}$, is a function that takes a weight vector $\mathbf{w}$ as input, and returns the maximal attainable scalarized value with any payoff vector in S:

$$V_{\mathcal{S}}^*(\mathbf{w}) = \max_{\mathbf{V}^\pi \in \mathcal{S}} \mathbf{w} \cdot \mathbf{V}^\pi.$$

If we take a complete CCS as $\mathcal{S}$, the scalarized value function $V_{CCS}^*(\mathbf{w})$ is the *optimal* scalarized value function. By definition, for every $\mathbf{w}$ the optimal scalarized value for the MODP is $V_{CCS}^*(\mathbf{w})$. In Figure 5.1(b), the optimal scalarized value function is represented by the bold line segments.

Similar to the scalarized value function, we define the set of maximizing value vectors and associated policies:

Definition 5.3   The optimal value vector set function with respect to $\mathcal{S}$ is a function that gives the value vectors that maximize the scalarized value for a given $\mathbf{w}$:

$$\mathcal{V}_{\mathcal{S}}(\mathbf{w}) = \operatorname*{argmax}_{\mathbf{V}^\pi \in \mathcal{S}} \mathbf{w} \cdot \mathbf{V}^\pi.$$

Similarly, the optimal policy set $\Pi_S(\mathbf{w})$ is the set of policies whose value vectors comprise $\mathcal{V}_S(\mathbf{w})$.

Note that $\mathcal{V}_S(\mathbf{w})$ and $\Pi_S(\mathbf{w})$ are sets because for some $\mathbf{w}$ there can be multiple value vectors that provide the same scalarized value.

Because $V_S^*(\mathbf{w})$ is the maximum scalarized value for each $\mathbf{w}$, it is the convex upper surface of all of these lines (which represent the value vectors in $\mathcal{S}$). Hence, $V_S^*(\mathbf{w})$ and $V_{CCS}^*(\mathbf{w})$ are *piecewise linear and convex* (PWLC) functions. The PWLC property can be exploited by outer loop methods to identify the CCS with perfect accuracy and in finite time, as we show in Section 5.3. Before doing so however—for those who might be familiar with partially observable decision problems—we describe the connection between multi-objective decision problems, and *partially observable Markov decision processes* (POMDPs).

5.2.1   THE RELATIONSHIP WITH POMDPS

The scalarized value function (Definition 5.3) resembles the optimal value function of a *partially observable Markov decision process* (POMDP) [Cheng, 1988, Kaelbling et al., 1998], which is also PWLC. The belief vectors b in POMDPs correspond to our weight vectors $\mathbf{w}$ and the α-vectors correspond to our value vectors $\mathbf{V}$. POMDPs and MODPs are thus related problems. This analogy, first identified by White and Kim [1980], is important because many techniques from the POMDP literature that exploit the PWLC property can also be exploited when computing CCSs for MODPs.

However, there are also important differences; while scalability in the number of states in a POMDP is key to the success of a POMDP solver, the number of objectives in an MODP is typically small—there are many MODPs with only two or three objectives. Therefore, scalability in the number of objectives is typically less important than scalability in other properties of an MODP. In the next section, we present an MODP method that builds off a POMDP method that has poor scalability in the number of POMDP states, but forms a good starting point for creating an MODP method.

5.3   OPTIMISTIC LINEAR SUPPORT

In this section, we present *optimistic linear support* (OLS) [Roijers, 2016, Roijers et al., 2014b, 2015b], a recent outer loop algorithm for computing the CCS for MODPs. Like RS, OLS deals with multiple objectives in the outer loop, by employing a single-objective method as a subroutine, and building the CCS incrementally. In each iteration, OLS adds at most one new vector to a partial CCS, $\mathcal{S}$. To find this vector, OLS selects a single linear scalarization weight, $\mathbf{w}$.

The key difference with RS is that OLS selects $\mathbf{w}$ intelligently. Specifically, OLS is optimistic: it selects the $\mathbf{w}$ that offers the maximal possible improvement—an upper bound on the difference between $V_S^*(\mathbf{w})$ and the optimal scalarized value function $V_{CCS}^*(\mathbf{w})$. After OLS identifies $\mathbf{w}$, it is passed to the inner loop, in which OLS calls a problem-specific single-objective solver to solve the single-objective decision problem that results from scalarizing the MODP using $\mathbf{w}$.

If the policy that is optimal for this scalarized problem is not already in the partial CCS, it is added to it, along with its value vector.

The departure point for creating OLS—*Cheng's linear support* (CLS) [Cheng, 1988]—was originally designed as a pruning algorithm for POMDPs. Unfortunately, CLS is rarely used for POMDPs in practice, as its runtime is exponential in the number of states. Scalability in the number of states is one of the major bottlenecks in this type of partially observable decision problem. However, the number of states in a POMDP corresponds to the number of objectives in an MODP, and while realistic POMDPs typically have many states, many MODPs have only a handful of objectives. Therefore, for MODPs, scalability in the number of objectives is typically less important than scalability in other properties of the problem (such as the number of agents or the number of states), making Cheng's linear support an attractive starting point for developing efficient MODP algorithms.

Because OLS takes an arbitrary single-objective problem solver as input, it can be seen as a generic multi-objective method that applies to any cooperative MODP. We show that OLS chooses a **w** at each iteration such that, after a finite number of iterations, no further improvements to the partial CCS can be made and OLS can terminate. Furthermore, we bound the maximum scalarized error of the intermediate results, so that they can be used as bounded approximations of the CCS. After defining the algorithm in this section, we show in Section 5.4 that OLS inherits any favorable properties from the single-objective subroutines it uses. When we instantiate OLS by using different single-objective solvers for different instances of MODPs, this leads to novel state-of-the-art algorithms for a variety of problems.

In this section, we assume that the single-objective solver used as a subroutine by OLS is exact, i.e., the scalarized instances are solved optimally. In other words, we assume that OLS has access to a function called `SolveSODP` that computes the best payoff vector for a given **w**. For now, we leave the implementation of `SolveSODP` abstract. As mentioned in Section 5.1, we can either use an adapted single-objective solver that returns the value vector, or we can employ a separate policy evaluation step. In Section 5.5, we relax the assumption that `SolveSODP` needs to be exact.

OLS exploits the PWLC property of the scalarized value function, $V_S^*(\mathbf{w})$ (Definition 5.2). OLS incrementally builds the CCS by adding to an initially empty partial CCS $\mathcal{S}$ new value vectors that are found at *corner weights*. These corner weights are the weights where $V_S^*(\mathbf{w})$ changes slope in all directions. These must thus be weights where $\mathcal{V}_S(\mathbf{w})$ and $\Pi_S(\mathbf{w})$ (Definition 5.3) consist of multiple value vectors and associated policies. Every corner weight is prioritized by the maximal possible improvement of finding a new value vector at that corner weight. When the maximal possible improvement is 0, OLS knows that the partial CCS is complete.

As an example of this process, we show the entire run of OLS for the problem of Table 4.1 in Figure 5.2. The corner weights where the algorithm has already searched for new value vectors are indicated by gray vertical lines; the corner weights that are still pending are indicated by red vertical lines.

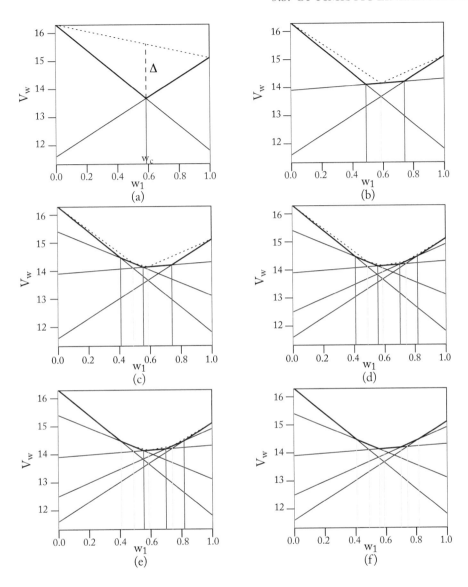

Figure 5.2: A run of OLS for the problem of Table 4.1. In each picture, V_S^* is indicated by bold line segments, corner weights are indicated by vertical lines (gray for visited, and red for pending), and $\overline{CCS}$ (which we define later in (Definition 5.6)) is shown as dotted lines. (a) OLS finds two payoff vectors at the extrema, and a new corner weight $\mathbf{w}_c = (0.4125, 0.5875)$ is found, with maximal possible improvement Δ. (b) OLS finds a new vector at $\mathbf{w}_c$, and adds two new corner weights to Q. (c,d) OLS finds another new vector and adds two new corner weights to Q. (e) OLS does not find a new vector for the corner weight it selects, reducing the maximal possible improvement at that corner weight to 0. (f) For the remaining corner weights, no new vectors are found, ensuring $S = \overline{CCS} = CCS$.

OLS is shown in Algorithm 5.8. OLS takes as input: m, the MODP to be solved, the single-objective subroutine, SolveSODP, and ε, the maximal tolerable error in the result.

Algorithm 5.8 OLS(m, SolveSODP, ε)

Input : A MODP m, a single-objective subroutine SolveSODP, and max. allowed error ε.
Output : An ε-CCS

1: $\mathcal{S} \leftarrow \emptyset$ //a partial CCS
2: $\mathcal{W} \leftarrow \emptyset$ //a set of visited weights
3: $Q \leftarrow$ an empty priority queue
4: **for all** extremum of the weight simplex $\mathbf{w}_e$ **do**
5:      Q.add($\mathbf{w}_e, \infty$) //add the extrema to Q with infinite priority
6: **end for**
7: **while** $\neg$Q.isEmpty() $\wedge$ $\neg$timeOut **do**
8:      $\mathbf{w} \leftarrow$ Q.pop()
9:      $\mathbf{V}^\pi \leftarrow$ SolveSODP($m, \mathbf{w}$)
10:     $\mathcal{W} \leftarrow \mathcal{W} \cup \{\mathbf{w}\}$
11:     **if** $\mathbf{V}^\pi \notin \mathcal{S}$ **then**
12:         $W_{del} \leftarrow$ remove the corner weights made obsolete by $\mathbf{V}$ from Q, and store them
13:         $W_{del} \leftarrow \{\mathbf{w}\} \cup W_{del}$ //corner weights to remove
14:         $W_{\mathbf{V}^\pi} \leftarrow$ newCornerWeights($\mathbf{V}^\pi, W_{del}, \mathcal{S}$)
15:
16:         $\mathcal{S} \leftarrow S \cup \{\mathbf{V}^\pi\}$
17:         **for all** $\mathbf{w} \in W_{\mathbf{V}^\pi}$ **do**
18:             $\Delta_r(\mathbf{w}) \leftarrow$ calculate improvement using maxValueLP($\mathbf{w}, \mathcal{S}, \mathcal{W}$)
19:
20:             **if** $\Delta_r(\mathbf{w}) > \varepsilon$ **then**
21:                 Q.add($\mathbf{w}, \Delta_r(\mathbf{w})$)
22:             **end if**
23:         **end for**
24:     **end if**
25: **end while**
26: **return** $\mathcal{S}$ and the highest $\Delta_r(\mathbf{w})$ left in Q

We first describe how OLS is initialized. Then, we define corner weights formally and describe how OLS identifies them. Finally, we describe how OLS prioritizes corner weights and how this can also be used to bound the error when stopping OLS before it is done finding a full CCS.

Initialization

OLS starts by initializing the partial CCS, $\mathcal{S}$, which contains the payoff vectors in the CCS that have been discovered so far (line 1 of Algorithm 5.8), as well as the set of visited weights $\mathcal{W}$ (line 2). Then, OLS adds the extrema of the weight simplex, i.e., those points where all of the weight is on one objective, to a priority queue Q, with infinite priority (line 5).

First Iterations

The extrema added during initialization are popped off the priority queue when OLS enters the main loop (line 7), in which the $\mathbf{w}$ with the highest priority is selected (line 8). Because the extrema have infinite priority, they are guaranteed to be popped off the queue first. SolveSODP is called with $\mathbf{w}$ (line 9) to find $\mathbf{V}$, the best payoff vector for that $\mathbf{w}$.

OLS assumes that SolveSODP returns the best value vector for $\mathbf{w}$. This can be done by adapting an out-of-the-box single-objective solver to return this value vector, or by adding a separate policy evaluation step. In the latter case, line 9 could instead be written as:

$\pi \leftarrow$ SolveSODP$(m, \mathbf{w})$
$\mathbf{V}^{\pi} \leftarrow$ PolicyEval(m, π)

Throughout this book, we assume that we can determine the exact value of $\mathbf{V}^{\pi}$ of the π that SolveSODP returns, i.e., the policy evaluation step is exact.

To illustrate the situation after initialization, Figure 5.2a shows $\mathcal{S}$ after two value vectors of the 2-objective problem of Table 4.1 have been found by applying SolveSODP to the extrema of the weight simplex. This leads to $S = \{(16.3, 11.8), (11.6, 15.1)\}$. Each of these vectors must be part of the CCS because it is optimal for at least one $\mathbf{w}$: the one for which SolveSODP returned it as a solution. The set of weights $\mathcal{W}$ that OLS has tested so far, i.e., the extrema of the weight simplex, are marked with vertical gray line segments.

Corner Weights

After having evaluated the extrema, $\mathcal{S}$ consists of d (the number of objectives) value vectors and associated policies. However, for many weights on the simplex, it does not yet contain the optimal value vector. Therefore, after identifying a new vector $\mathbf{V}^{\pi}$ to add to S, OLS must determine which new weight vectors to add to Q, and with what priority. Like Cheng's linear support, OLS identifies the *corner weights*: the weights at the corners of the convex upper surface, i.e., the points where the PWLC surface $V_{S}^{*}(\mathbf{w})$ changes slope. To define the corner weights precisely, we define P, the polyhedral subspace that is above $V_{S}^{*}(\mathbf{w})$ [Bertsimas and Tsitsiklis, 1997]. For example, in Figure 5.1 (b), P is displayed as the shaded area above the optimal scalarized value function. The corner weights are the vertices of P, which can be defined by a set of linear inequalities:

Definition 5.4   If S is the set of known payoff vectors, we define a polyhedron

$$P = \{\mathbf{x} \in \Re^{d+1} : \mathcal{S}^{+}\mathbf{x} \geq \vec{0}, \forall i, w_i > 0, \sum_{i} w_i = 1\},$$

Algorithm 5.9 newCornerWeights($\mathbf{V}^{\pi}_{new}, W_{del}, \mathcal{S}$)

Input : A new value vector, $\mathbf{V}^{\pi}_{new}$, a set of obsolete corner weights, W_{del}, and the current partial CCS, $\mathcal{S}$

Output : A set of new corner weights

1: $\mathcal{V}_{rel} \leftarrow \bigcup_{\mathbf{w} \in W_{del}} \mathcal{V}_{\mathcal{S}}(\mathbf{w})$ //involved value vectors (see Definition 5.3)

2: $\mathcal{B}_{rel} \leftarrow$ the set of boundaries of the weight simplex involved in any $\mathbf{w} \in W_{del}$;

3: $W_{new} \leftarrow \emptyset$ //new corner weights

4: **for all** subset X of $d - 1$ elements from $\mathcal{V}_{rel} \cup \mathcal{B}_{rel}$ **do**

5: $\quad \mathbf{w}_c \leftarrow$ compute weight where the vectors/boundaries in X intersect with $\mathbf{V}^{\pi}_{new}$;

6: $\quad$ **if** $\mathbf{w}_c$ is inside the weight simplex **then**

7: $\qquad$ **if** $\mathbf{w}_c \cdot \mathbf{V}^{\pi}_{new} = V^*_{\mathcal{S}}(\mathbf{w}_c)$ **then**

8: $\qquad\quad W_{new} \leftarrow W_{new} \cup \mathbf{w}_c$

9: $\qquad$ **end if**

10: $\quad$ **end if**

11: **end for**

12: **return** W_{new}

---

where $\mathcal{S}^+$ is a matrix with the elements of S as row vectors, augmented by a column vector of -1's. The set of linear inequalities, $\mathcal{S}^+\mathbf{x} \geq \vec{0}$, is supplemented by the simplex constraints: $\forall i \ w_i > 0$ and $\sum_i w_i = 1$. The vector $\mathbf{x} = (w_1, \ldots, w_d, V_{\mathbf{w}})$ consists of a weight vector and a scalarized value at those weights. The *corner weights* are the weights contained in the vertices of P, which are also of the form $(w_1, \ldots, w_d, V_{\mathbf{w}})$.

Note that, due to the simplex constraints, P is only d-dimensional. Furthermore, the extrema of the weight simplex are special cases of corner weights.

After identifying the new value vector $\mathbf{V}^{\pi}$, OLS identifies which corner weights change in the polyhedron P by adding $\mathbf{V}^{\pi}$ to S. Fortunately, this does not require re-computation of all the corner weights, but can be done incrementally: first, the corner weights in Q for which $\mathbf{V}^{\pi}$ yields a better scalarized value than currently known are deleted from the queue (line 12) and then the function newCornerWeights($\mathbf{V}^{\pi}, W_{del}, \mathcal{S}$) (line 14) calculates the new corner weights that involve $\mathbf{V}^{\pi}$.

The function newCornerWeights($\mathbf{V}^{\pi}, W_{del}, \mathcal{S}$) (Algorithm 5.9) first calculates the set of all relevant payoff vectors, $\mathcal{V}_{rel}$, by taking the union of all the maximizing vectors of the weights in W_{del} for $\mathcal{S}$ (on line 1). In the implementation, we can optimize this step by caching $\mathcal{V}_{\mathcal{S}}(\mathbf{w})$ (Definition 5.3) and associated policies for all $\mathbf{w}$ in Q. If for a corner weight, $\mathbf{w}$, $\mathcal{V}_{\mathcal{S}}(\mathbf{w})$ contains fewer than d value vectors, then a boundary of the weight simplex is involved. These boundaries are also stored (line 2). All possible subsets of size $d{-}1$—of both vectors and boundaries—

are taken. For each subset, the weight where these $d - 1$ payoff vectors (and boundaries) intersect with each other and $\mathbf{V}$ is computed by solving a system of linear equations (on line 5). The intersection weights for all subsets together form the set of candidate corner weights: W_{can}. newCornerWeights($\mathbf{V}^\pi, W_{del}, S$) returns the subset of W_{can} that is inside the weight simplex and for which there is no vector $\mathbf{V}' \in S$ that has a higher scalarized value than $\mathbf{V}$ at that weight. OLS then adds the new corner weights returned by newCornerWeights($\mathbf{V}^\pi, W_{del}, S$) to its queue, Q.

In Figure 5.2a, after computing the value vectors for the two extrema, one new corner weight is produced, labeled $\mathbf{w}_c = (0.4125, 0.5875)$. In the subsequent iterations where a new value vector is identified (Figure 5.2b–d), two new corner weights are produced. For two objectives, there are always (at most) two new corner weights per iteration, and two value vectors involved in each corner weights. For higher numbers of objectives, it is in theory possible to construct a partial CCS, S that has a corner weight for which all payoff vectors in S are in $\mathcal{V}_{rel}$, leading to very many new corner weights. In practice however, $|\mathcal{V}_{rel}|$ is typically small, and only a few systems of linear equations need to be solved, leading to a limited number of new corner weights.

After calculating the new corner weights $W_{\mathbf{V}^\pi}$ at line 14 (in Algorithm 5.8), $\mathbf{V}$ is added to S at line 16. Cheng showed that finding the best payoff vector for each corner weight and adding it to the partial CCS guarantees the best improvement to S:

Theorem 5.5   *[Cheng, 1988] The maximum value of:*

$$\max_{\mathbf{w}, \mathbf{V} \in CCS} \min_{\mathbf{V}' \in S} \mathbf{w} \cdot \mathbf{V} - \mathbf{w} \cdot \mathbf{V}',$$

i.e., the maximal improvement to S by adding a vector to it, is at one of the corner weights.

Theorem 5.5 guarantees the correctness of OLS: after all corner weights are checked, there are no new payoff vectors; thus the maximal improvement must be 0 and OLS has found the full CCS (as is the case in Figure 5.2f).

Prioritization

Cheng's linear support assumes that all corner weights can be checked inexpensively, which is a reasonable assumption in a POMDP setting. However, since SolveSODP is typically an expensive operation, testing all corner weights is usually infeasible in MODPs. For example, in MO-CoGs, a common choice for SolveSODP is the *variable elimination* algorithm (Section 4.2.1), whose runtime can be exponential in the size of the problem. Therefore, unlike Cheng's linear support, OLS pops only one $\mathbf{w}$ off Q to be tested per iteration. Making OLS efficient thus critically depends on giving each $\mathbf{w}$ a suitable priority when adding it to Q. To this end, OLS prioritizes each corner weight $\mathbf{w}$ according to its *maximal possible improvement*, an upper bound on the improvement to $V_S^*(\mathbf{w})$ that can be made by adding a single new value vector. This upper bound is computed with respect to $\overline{CCS}$, the *optimistic hypothetical CCS*, i.e., the best-case scenario for the final CCS given that $\mathcal{S}$ is the current partial CCS and $\mathcal{W}$ is the set of weights already tested with SolveSODP. A key advantage of OLS over Cheng's linear support is thus that these priorities can be computed without calling SolveSODP, obviating the need to run SolveSODP on all corner weights.

Definition 5.6   An *optimistic hypothetical CCS*, $\overline{CCS}$, is a set of payoff vectors that yields the highest possible scalarized value for all possible $\mathbf{w}$ consistent with finding the vectors $\mathcal{S}$ at the weights in $\mathcal{W}$.

In Figure 5.2a the $\overline{CCS} = \{(16.3, 11.8), (11.6, 15.1), (16.3, 15.1)\}$. $\overline{CCS}$ is a superset of $\mathcal{S}$ and the value of $V^*_{\overline{CCS}}(\mathbf{w})$ (indicated by the dotted line) is the same as $V^*_{\mathcal{S}}(\mathbf{w})$ at all the weights in $\mathcal{W}$. For a given $\mathbf{w}$, maxValueLP($\mathbf{w}, \mathcal{S}, \mathcal{W}$) finds the scalarized value of $V^*_{\overline{CCS}}(\mathbf{w})$ by solving the following linear program:

$$\max \ \mathbf{w} \cdot \mathbf{v}$$
$$\text{subject to} \quad \mathcal{W}\mathbf{v} \leq \mathbf{V}^*_{\mathcal{S},\mathcal{W}},$$

where $\mathbf{V}^*_{\mathcal{S},\mathcal{W}}$ is a vector containing $V^*_{\mathcal{S}}(\mathbf{w}')$ for all $\mathbf{w}' \in \mathcal{W}$, and $\mathbf{v}$ is a vector of variables of length d. Note that we abuse the notation $\mathcal{W}$, which in this case is a matrix whose rows correspond to all the weight vectors in the set $\mathcal{W}$.[1]

Using $\overline{CCS}$, we can define the maximal possible improvement of each $\mathbf{w}$:

$$\Delta(\mathbf{w}) = V^*_{\overline{CCS}}(\mathbf{w}) - V^*_{\mathcal{S}}(\mathbf{w}).$$

Figure 5.2a shows $\Delta(\mathbf{w}_c)$ with a dashed line. We use the *maximal relative possible improvement*, $\Delta_r(\mathbf{w}) = \Delta(\mathbf{w})/V^*_{\overline{CCS}}(\mathbf{w})$, as the priority of each new corner weight $\mathbf{w} \in W_{\mathbf{V}}$. In Figure 5.2a, $\Delta_r(\mathbf{w}_c) = \frac{(0.4125, 0.5875) \cdot ((16.3, 15.1) - (11.6, 15.1))}{13.65625} = 0.141968$. When a corner weight $\mathbf{w}$ is identified (line 14), it is added to Q with priority $\Delta_r(\mathbf{w})$ as long as $\Delta_r(\mathbf{w}) > \varepsilon$ (lines 18–21). In Figure 5.2 the corner weights in Q are indicated by red vertical line segments.

After $\mathbf{w}_c$ in Figure 5.2a is added to Q, it is popped off again (as it is the only element of Q). SolveSODP($\mathbf{w}_c$) generates a new value vector $(13.9, 14.3)$, yielding $\mathcal{S} = \{(16.3, 11.8), (11.6, 15.1), (13.9, 14.3)\}$, as illustrated in Figure 5.2b. The new corner weights are the points at which $(13.9, 14.3)$ intersects with $(16.3, 11.8)$ and $(11.6, 15.1)$. Testing these weights, as illustrated in Figure 5.2cd, results in 2 new payoff vectors, and two new corner weights each. After these two value vectors however, checking the weight with the highest priority in Q does not result in a new vector, reducing the maximal possible improvement at that weight to 0 (Figure 5.2e). Checking the remaining three corner weights in Q also does not lead to new value vectors, causing OLS to terminate. Because the maximal improvement at these corner weights is 0 upon termination, $S = CCS$ due to Theorem 5.5. OLS called SolveSODP for only nine weights resulting exactly in the five payoff vectors of the CCS. The other seven payoff vectors in $\mathcal{V}$ (the black points of Figure 5.1) were never generated.

---

[1]Our implementation of OLS reduces the size of the LP by using only the subset of weights in $\mathcal{W}$ for which the policies involved in $\mathbf{w}$, $\Pi_{\mathcal{S}}(\mathbf{w})$, have been found to be optimal. This can lead to a slight overestimation of $V^*_{\overline{CCS}}(\mathbf{w})$.

5.4   ANALYSIS

We now analyze the computational and space complexity of OLS. Because OLS is a generic algorithm that takes a single-objective solver (and possibly a policy evaluation algorithm) as a subroutine, we provide the complexity bounds of OLS in terms of the runtime and space complexities of these subroutines. We denote the runtime of a single-objective solver as R_{so} and the runtime of policy evaluation as R_{pe}, and the corresponding memory as M_{so} and M_{pe}.

Theorem 5.7   *The runtime of OLS is*

$$O(\,(|\varepsilon\text{–}CCS| + |\mathcal{W}_{\varepsilon\text{–}CCS}|)(R_{so} + R_{pe} + R_{nw} + R_{heur})\,),$$

where $|\varepsilon\text{–}CCS|$ is the size of the ε-CCS (Definition 3.16) outputted by OLS, $|\mathcal{W}_{\varepsilon\text{–}CCS}|$ is the number of corner weights of the scalarized $V^*_{\varepsilon\text{–}CCS}(\mathbf{w})$ corresponding to the output ε-CCS, R_{nw} the time it takes to run `newCornerWeights`, *and R_{heur} the time it takes to compute the value of the optimistic CCS using* `maxValueLP`.

Proof.   The runtime of one iteration of OLS is the cost of running the single-objective solver plus policy evaluation, $R_{so} + R_{pe}$, plus the overhead per corner weight $R_{nw} + R_{heur}$, multiplied by the number iterations. To count the number of iterations, we consider two cases: calls to the single-objective solver that result in adding a new vector to the partial CCS, $\mathcal{S}$ and those that do not result in a new vector but instead confirm the optimality of the scalarized value $V^*_{\mathcal{S}}(\mathbf{w})$ at that weight. The former is the size of the output of OLS, i.e., $|\varepsilon\text{–}CCS|$, while the latter is at most the number of corner weights of the scalarized value function of that same output set, $|\mathcal{W}_{\varepsilon\text{–}CCS}|$.   □

Note that we can often adapt the implementation of `SolveSODP` to return the value vector of the optimal policy for a weight, alongside this optimal policy, without an increase in the complexity bounds of the single-objective solver, in which case $R_{pe} = 0$.

In practice, the overhead of OLS, i.e., computing new corner weights, R_{nw}, and calculating the maximal relative improvement, R_{heur}, is small compared to the `SolveSODP` for the decision problems considered in this book. Typically, `newCornerWeights(u, `W_{del}`, S)` computes the solutions to only a small set of linear equations (of d equations each). `maxValueLP(w, S, W)` computes the solutions to linear programs, whose costs is polynomial in the size of its inputs.[2]

If $\varepsilon \neq 0$, OLS does not in practice test all corner weights of the polyhedron spanned by the ε-CCS it outputs, as it only tests a corner weight if the maximal possible improvement is larger than ε. However, this cannot be guaranteed in general. Note that if $\varepsilon = 0$, OLS outputs an exact CCS, and must check every corner weight in order to verify that it has done so correctly.

For $d = 2$, the number of corner weights is smaller than $|\varepsilon\text{–}CCS|$, the size of the output of OLS. Therefore, the runtime of OLS is $O(|\varepsilon\text{–}CCS|(R_{so} + R_{pe} + R_{nw} + R_{heur}))$. For $d = 3$, the number of corner weights is $2|\varepsilon\text{–}CCS|$ (minus a constant) because, when `SolveSODP` finds

---

[2]When the reduction in Footnote 1 is used, only a small subset of $\mathcal{W}$ is used, making it even cheaper.

a new payoff vector, one corner weight is removed and three new corner weights are added. For $d = 3$, the computational complexity is thus still only $O(|\varepsilon\text{-}CCS|(R_{so} + R_{pe} + R_{nw} + R_{heur}))$. For $d > 3$, a loose bound on $|\mathcal{W}_{\varepsilon\text{-}CCS}|$ is the total number of possible combinations of d payoff vectors or boundaries: $O(\binom{|\varepsilon\text{-}CCS|+d}{d})$. However, we can obtain a tighter bound by observing that counting the number of corner weights given a CCS is equivalent to *vertex enumeration*, which is the dual problem of *facet enumeration*, i.e., counting the number of vertices given the corner weights [Kaibel and Pfetsch, 2003].

Theorem 5.8   *[Avis and Devroye, 2000] For arbitrary d, $|\mathcal{W}_{\varepsilon\text{-}CCS}|$ is bounded by*

$$O\left(\binom{|\varepsilon\text{-}CCS| - \lfloor\frac{d+1}{2}\rfloor}{|\varepsilon\text{-}CCS| - d} + \binom{|\varepsilon\text{-}CCS| - \lfloor\frac{d+2}{2}\rfloor}{|\varepsilon\text{-}CCS| - d}\right).$$

Proof. This result follows directly from *McMullen's upper bound theorem* for facet enumeration [Henk et al., 1997, McMullen, 1970]. ☐

Theorem 5.9   *The space complexity of OLS is*

$$O\left(d\,|\varepsilon\text{-}CCS| + d\,|\mathcal{W}_{\varepsilon\text{-}CCS}| + M_{so} + M_{pe}\right).$$

Proof. OLS needs to store every corner weight (a vector of length d) in the queue, which is at most $|\mathcal{W}_{\varepsilon\text{-}CCS}|$. OLS also needs to store every vector in $\mathcal{S}$ (also vectors of length d). Furthermore, when SolveSODP is called, the memory usage of this single-objective solver is added to the memory usage of the outer loop of OLS. The same holds for policy evaluation. ☐

Because OLS adds few memory requirements to that of the single-objective solver for small and medium numbers of objectives, OLS is almost as memory efficient as the single-objective solver itself in these cases. By contrast, inner loop methods (Chapter 4) are less space efficient because they must retain a set of value vectors and partial policies in the place of each single value and single partial policy retained by the corresponding single-objective method.

5.5   APPROXIMATE SINGLE-OBJECTIVE SOLVERS

So far, we assumed that OLS takes an arbitrary *exact* single-objective solver as a subroutine. In this section, we show that OLS can also be applied when the single-objective subroutine is approximate. Furthermore, when the subroutine produces bounded approximations with at most ε error, OLS is guaranteed to produce an ε-CCS [Roijers, 2016, Roijers et al., 2014a].

First, we define an ε-approximate single-objective subroutine.

Definition 5.10   An ε-*approximate SODP solver* is an algorithm that produces a policy whose value is at least $(1 - \varepsilon)V^*$, where V^* is the optimal value for the SODP and $\varepsilon \geq 0$.

Given an ε-approximate SODP solver, we can compute a set of policies for which the scalarized value for each possible $\mathbf{w}$ is at least $1 - \varepsilon$ times the optimal scalarized value, i.e., an ε-*CCS* (Definition 3.16). To this end, we change OLS to handle approximate solvers, yielding *approximate OLS* (AOLS) (Algorithm 5.10). We now discuss the differences with OLS (Algorithm 5.8), which are highlighted in blue.

Firstly, Algorithm 5.10 assumes the single-objective subroutine approxSolveSODP returns both the value vector and policy,[3] $\mathbf{V}^\pi$, as well as an upper bound on the scalarized value at the weight $\mathbf{w}$ for which it is called, $\bar{V}_\mathbf{w}$ (line 9). This upper bound is stored alongside $\mathbf{w}$ (lines 2 and 10), and used instead of the scalarized values in maxValueLP. Specifically, for a given $\mathbf{w}$, maxValueLP finds (line 18) the scalarized value of $V^*_{\overline{CCS}}(\mathbf{w})$ by solving:

$$\begin{aligned} \max \quad & \mathbf{w} \cdot \mathbf{v} \\ \text{subject to} \quad & \mathcal{W}' \, \mathbf{v} \leq \bar{\mathbf{V}}^*_{S,\mathcal{W}}, \end{aligned}$$

where for each tuple $(\mathbf{w}, \bar{V}_\mathbf{w}) \in \mathcal{W}$, there is a row in the matrix $\mathcal{W}'$ corresponding to $\mathbf{w}$ with a corresponding element $\bar{V}_\mathbf{w}$ in the vector $\bar{\mathbf{V}}^*_{S,\mathcal{W}}$.

Secondly, because the single-objective subroutine is approximate, the vectors in the partial CCS S may no longer be optimal for any $\mathbf{w}$ once we add a new value vector. Therefore, we must first remove these vectors from S (line 15). We can do this efficiently by checking whether $\mathbf{V}^\pi$ is better than a vector $\mathbf{V}' \in S$ for all the corner weights of $V^*_S(\mathbf{w})$ for which $\mathbf{V}'$ is optimal, as this bounds the area for which $\mathbf{V}'$ is optimal with respect to the other value vectors currently in S.

Correctness

We now establish the correctness of Algorithm 5.10, i.e., OLS with approximate single-objective subroutines. Because the scalarized value of $V^*_{\overline{CCS}}(\mathbf{w})$ (as computed by maxValueLP), obtained using the approximate solveMDP, is no longer identical to $V^*_S(\mathbf{w})$, we need to modify Cheng's theorem. The following theorem is identical to Cheng's, except that S is no longer a subset of the CCS, necessitating a new proof.

Theorem 5.11   *There is a corner weight of $V^*_S(\mathbf{w})$ that maximizes:*

$$\Delta(\mathbf{w}) = V^*_{CCS}(\mathbf{w}) - V^*_S(\mathbf{w}),$$

where S is an intermediate set of value vectors computed by AOLS (Algorithm 5.10).

---

[3]We assume that the value $\mathbf{V}^\pi$ is the correct value vector of π, i.e., the evaluation of the policy value is exact.

Algorithm 5.10 AOLS(m, approxSolveSODP, ε)

Input : An MODP m, an approximate SODP subroutine ApproxSolveSODP, and max. allowed error, ε. *Output* : An ε-CCS

1: $\mathcal{S} \leftarrow \emptyset$ //a partial CCS
2: $\mathcal{W} \leftarrow \emptyset$ //set of tuples of visited weights and upper bounds, $\bar{V}_{\mathbf{w}}$
3: $Q \leftarrow$ an empty priority queue
4: **for all** extremum of the weight simplex $\mathbf{w}_e$ **do**
5:    Q.add($\mathbf{w}_e, \infty$) //add the extrema to Q with infinite priority
6: **end for**
7: **while** $\neg Q$.isEmpty() $\wedge \neg$timeOut **do**
8:    $\mathbf{w} \leftarrow Q$.pop()
9:    $\mathbf{V}^\pi, \bar{V}_{\mathbf{w}} \leftarrow$ approxSolveSODP($m, \mathbf{w}$)
10:    $\mathcal{W} \leftarrow \mathcal{W} \cup \{(\mathbf{w}, \bar{V}_{\mathbf{w}})\}$
11:    **if** $\mathbf{V}^\pi \notin \mathcal{S}$ **then**
12:       $W_{del} \leftarrow$ remove the corner weights made obsolete by $\mathbf{V}$ from Q, and store them
13:       $W_{del} \leftarrow \{\mathbf{w}\} \cup W_{del}$ //corner weights to remove
14:       $W_{\mathbf{V}^\pi} \leftarrow$ newCornerWeights($\mathbf{V}^\pi, W_{del}, \mathcal{S}$)
15:       remove vectors from $\mathcal{S}$ that are no longer optimal for any $\mathbf{w}$ when $\mathbf{V}^\pi$ is added
16:       $\mathcal{S} \leftarrow (S \cup \{\mathbf{V}^\pi\})$
17:       **for all** $\mathbf{w} \in W_{\mathbf{V}^\pi}$ **do**
18:          $\Delta_r(\mathbf{w}) \leftarrow$ calculate improvement using maxValueLP($\mathbf{w}, \mathcal{S}, \mathcal{W}$)
19:          **if** $\Delta_r(\mathbf{w}) > \varepsilon$ **then**
20:             Q.add($\mathbf{w}, \Delta_r(\mathbf{w})$)
21:          **end if**
22:       **end for**
23:    **end if**
24: **end while**
25: **return** $\mathcal{S}$ and the highest $\Delta_r(\mathbf{w})$ left in Q

Proof. $\Delta(\mathbf{w})$ is the difference between two PWLC functions: $V_{CCS}^*(\mathbf{w})$ and $V_{\mathcal{S}}^*(\mathbf{w})$. To maximize this function, we have three possible cases, shown in Figure 5.3: (a) the maximum is at a weight that is neither a corner point of $V_{CCS}^*(\mathbf{w})$, nor of $V_{\mathcal{S}}^*(\mathbf{w})$; (b) it is at a corner point of $V_{CCS}^*(\mathbf{w})$ but not of $V_{\mathcal{S}}^*(\mathbf{w})$; or (c) it is at a corner point of $V_{\mathcal{S}}^*(\mathbf{w})$.

Case (a) can only apply if the slope of $\Delta(\mathbf{w})$ at $\mathbf{w}$ is 0. If this is so, then the value of $\Delta(\mathbf{w})$ is equal to the value at the corner points where this slope changes. Case (b) can never occur: if $\mathbf{w}$ is a corner point of $V_{CCS}^*(\mathbf{w})$, and not of $V_{\mathcal{S}}^*(\mathbf{w})$, and $\Delta(\mathbf{w}) = V_{CCS}^*(\mathbf{w}) - V_{\mathcal{S}}^*(\mathbf{w})$ is at a maximum, then, because $V_{\mathcal{S}}^*(\mathbf{w})$ does not change slope in $\mathbf{w}$, the change in slope for $V_{CCS}^*(\mathbf{w})$

must be negative. However, because we know that $V^*_{CCS}(\mathbf{w})$ is a PWLC function, this leads to a contradiction. Therefore, only case (c) remains. $\square$

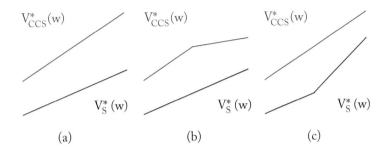

Figure 5.3: Possible cases for the maximum possible improvement of S with respect to the CCS.

Because the maximum possible improvement is still at the corner points of S, even though S now contains ε-approximate solutions, the original scheme of calling a single-objective subroutine for the corner weights still applies.

The approximate-subroutine version of OLS terminates when the maximal possible improvement is less than or equal to $\varepsilon \cdot V^*_{CCS}(\mathbf{w})$. In other words, when there are no corner weights with a possible improvement higher than the input slack ε.

Theorem 5.12   *AOLS (Algorithm 5.10) terminates after a finite number of calls to an ε-approximate SODP solver* `approxSolveSODP` *and produces an ε-CCS.*

Proof. AOLS runs until there are no corner points left in the priority queue to check, and returns S and the highest priority left in Q. Once a corner point is evaluated, it is never considered again because the established value lies within $(1 - \varepsilon)V^*_{CCS}(\mathbf{w})$ (as guaranteed by the ε-bound of `approxSolveSODP`). AOLS thus terminates after checking a finite number of corner weights. All other corner weights have a possible improvement less than or equal to $\varepsilon V^*_{CCS}(\mathbf{w})$. Therefore, S must be an ε-CCS. $\square$

An equivalent, but subtly different version of the approximate OLS algorithm, and the corresponding correctness theorem, can be established accordingly, when the single-objective subroutine provides an upper and lower bound (implying an ε), but we do not know beforehand how strict these bounds will be [Roijers et al., 2015a].

Corollary 5.13   *When an approximate single-objective solver produces a bounded approximate solution for each scalarized problem, with an error bound of at most ε, AOLS produces an ε-CCS.*

Proof. The proof is the same as for Theorem 5.12. We assume that there is some ε for which `approxSolveSODP` is ε-approximate and establish its value on the fly by inspecting the difference between the upper and lower bounds found by the ε-approximate solver. $\square$

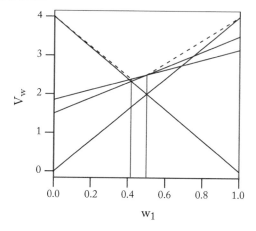

Figure 5.4: Close corner weights lead to close value vectors.

Note that the runtime and memory requirements, as established in Theorems 5.7 and 5.9, are not affected by the use of an approximate single-objective subroutine instead of an exact one, except perhaps changing the values of R_{so} and M_{so}.

5.6   VALUE REUSE

In OLS, the single-objective subroutine is called for every corner weight **w**. When there are many corner weights, this may be prohibitively expensive. In this section, we discuss an extension to OLS that addresses this issue. Rather than each call to the single-objective solver starting from scratch, we *reuse* part of the work done in earlier iterations to *hot-start* the single-objective subroutine for a new **w** [Roijers, 2016, Roijers et al., 2015a,c, Van Doorn et al., 2016].

The key insight behind *reuse in OLS*, is that when two corner weights **w** and **w**′ are similar, the value vectors for these weights found by the single-objective solver are also likely to be similar (e.g., as in Figure 5.4). Therefore, we also expect that similar policies are optimal for **w**′. If that is indeed the case, and we have a single-objective subroutine that can start from a previous solution—or part of a previous solution—and gradually improve, this may save a lot of runtime.

Algorithm 5.11 implements this idea. The differences with Algorithm 5.10 are shown in blue. We aim to show the most general form of the algorithm, and therefore use an abstract data type $I_{\mathbf{w}}$ that represents *any possible information* from a previous call to approxSolveSODP for a weight **w** that we may reuse in subsequent iterations. Which information can be reused depends on the specifics of the MODP, e.g., the full state-value function of an MOMDP. We assume that this information is produced by the single-objective solver (line 10), and stored together with the previous search weights (lines 2 and 11). At every iteration, all possibly reusable information from

Algorithm 5.11 OLS+R(m, approxSolveSODP, ε)

Input: An MODP m, an approximate SODP subroutine ApproxSolveSODP, and max. allowed error, ε. *Output*: An ε-CCS

1: $\mathcal{S} \leftarrow \emptyset$ //a partial CCS
2: $\mathcal{W} \leftarrow \emptyset$ //a set of tuples of $\mathbf{w}$, $\bar{V}_\mathbf{w}$ and $I_\mathbf{w}$
3: $Q \leftarrow$ an empty priority queue;
4: **for all** extremum of the weight simplex $\mathbf{w}_e$ **do**
5:     Q.add($\mathbf{w}_e$, ∞) //add the extrema to Q with infinite priority
6: **end for**
7: **while** $\neg Q$.isEmpty() $\wedge$ $\neg$timeOut **do**
8:     $\mathbf{w} \leftarrow Q$.pop();
9:     $I \leftarrow$ Retrieve the reusable information for relevant previous iterations from $\mathcal{W}$;
10:     $\mathbf{V}^\pi, \bar{V}_\mathbf{w}, I_\mathbf{w} \leftarrow$ approxSolveSODP($m, \mathbf{w}, I$);
11:     $\mathcal{W} \leftarrow \mathcal{W} \cup \{(\mathbf{w}, \bar{V}_\mathbf{w}, I_\mathbf{w})\}$;
12:     **if** $\mathbf{V}^\pi \notin \mathcal{S}$ **then**
13:         $W_{del} \leftarrow$ remove the corner weights made obsolete by $\mathbf{V}$ from Q, and store them
14:         $W_{del} \leftarrow \{\mathbf{w}\} \cup W_{del}$ //corner weights to remove
15:         $W_{\mathbf{V}^\pi} \leftarrow$ newCornerWeights($\mathbf{V}^\pi, W_{del}, \mathcal{S}$);
16:         remove vectors from $\mathcal{S}$ that are no longer optimal for any $\mathbf{w}$ when $\mathbf{V}^\pi$ is added;
17:         $\mathcal{S} \leftarrow (\mathcal{S} \cup \{\mathbf{V}^\pi\})$;
18:         **for all** $\mathbf{w} \in W_{\mathbf{V}^\pi}$ **do**
19:             $\Delta_r(\mathbf{w}) \leftarrow$ calculate improvement using maxValueLP($\mathbf{w}, \mathcal{S}, \mathcal{W}$);
20:             **if** $\Delta_r(\mathbf{w}) > \varepsilon$ **then**
21:                 Q.add($\mathbf{w}, \Delta_r(\mathbf{w})$);
22:             **end if**
23:         **end for**
24:     **end if**
25: **end while**
26: **return** $\mathcal{S}$ and the highest $\Delta_r(\mathbf{w})$ left in Q

---

all previous iterations, I, is retrieved (line 9) and used to hot-start the single-objective subroutine on line 10.

Without further assumptions, *reuse* is a heuristic, i.e., it may improve OLS's runtime in practice, but it does not improve its complexity. However, if we assume that I contains enough information for approxSolveSODP to check that $\mathcal{S}$ is already sufficient for the new weight $\mathbf{w}$, and that this check can be performed in $R_{confirm}$, we do not require a full run of the single-objective subroutine for corner weights for which we do not find a new value vector, and can tighten the theoretical runtime guarantees.

Theorem 5.14   *The runtime of OLS with reuse is*

$$O(\,(|\varepsilon\text{–}CCS|)(R_{so} + R_{pe} + R_{nw} + R_{heur}) + |\mathcal{W}_{\varepsilon-CCS}|R_{confirm}\,),$$

where $R_{confirm}$ is the time it takes to check the sufficiency of the solutions currently in $\mathcal{W}$ and their value vectors $\mathcal{S}$.

This is especially important when the number of objectives is high, because, as Theorem 5.8 shows, $|\mathcal{W}_{\varepsilon-CCS}|$ increases exponentially in the number of objectives. However, the more information stored per $\mathbf{w}$, the more memory per corner weight OLS with reuse uses.

5.7   COMPARING AN INNER AND OUTER LOOP METHOD

In this section, we make a theoretical and empirical comparison between an inner and outer loop method. To make these comparisons fair, we build both methods from the same single-objective method, namely *variable elimination* (VE) for CoGs (Section 4.2.1).

In the previous chapter, we used VE to create the inner loop CMOVE algorithm for CCS planning in MO-CoGs (Section 4.2.3). We can also take an outer loop approach and employ VE as a subroutine in OLS, yielding a method we call *variable elimination linear support* (VELS) [Roijers et al., 2014b, 2015b].

VELS is implemented as OLS(m, SolveSODP, ε) (Algorithm 5.8), where m is a MO-CoG and SolveSODP consists of two steps: first calling VE to retrieve the optimal action, $\mathbf{a}$ for a scalarized instance of m, and then evaluating $\mathbf{a}$ to retrieve the payoff vector $\mathbf{u}(\mathbf{a})$.

5.7.1   THEORETICAL COMPARISON

The computational and space complexities of VELS can be immediately derived from the complexities of OLS (Theorems 5.7 and 5.9) VE (Theorems 4.5 and 4.6).

Corollary 5.15   *The computational complexity of VELS*

$$O(\,(|\varepsilon\text{–}CCS| + |\mathcal{W}_{\varepsilon-CCS}|)\,(n|\mathcal{A}_{max}|^{w} + d\rho + R_{nw} + R_{heur})\,),$$

where $|\mathcal{A}_{max}|$ is the maximal number of actions for a single agent *and w is the induced width, ρ is the number of factors, and R_{nw} and R_{heur} are the time it costs to run* newCornerWeights, *and* maxValueLP, *as defined in Section 5.3.*

Proof. This follows directly from filling in the runtime of VE, $n|\mathcal{A}_{max}|^{w}$, for R_{so}, and the look-up and summation of ρ local payoff vectors of length d for R_{pe}, in Theorem 5.7.   □

The overhead of OLS itself, i.e., computing new corner weights, R_{nw}, and calculating the maximal relative improvement, R_{heur}, is small compared to the VE calls. $R_{nw} + R_{heur}$ and $d\rho$ are thus negligible in practice.

In CMOVE, the runtime of VE is multiplied by the runtime of the pruning operators (R_1 and R_2 in Theorem 4.13), while in VELS, the runtime of VE is multiplied by ($|\varepsilon\text{–}CCS| + |\mathcal{W}_{\varepsilon\text{–}CCS}|$). Considering that $|\mathcal{W}_{\varepsilon\text{–}CCS}|$ is linear in $|\varepsilon\text{–}CCS|$ for $d = 2$ and $d = 3$, while CPrune is polynomial in the size of the local CCSs, we thus expect VELS to be much faster for these numbers of objectives. However, because $|\mathcal{W}_{\varepsilon\text{–}CCS}|$ grows exponentially with the number of objectives (Theorem 5.8), we expect CMOVE to be faster than VELS for larger numbers of objectives.

Corollary 5.16 *The space complexity of VELS is*

$$O\left(d\,|\varepsilon\text{–}CCS| + d\,|\mathcal{W}_{\varepsilon\text{–}CCS}| + n\,|\mathcal{A}_{max}|^w\right).$$

Proof. This follows directly from filling in the memory requirements of VE, $n|\mathcal{A}_{max}|^w$, for M_{so}, in Theorem 5.7. The memory requirements for computing the payoff vector, $\mathbf{u}(\mathbf{a})$, given $\mathbf{a}$ are negligible. □

Since OLS adds few memory requirements to those of VE, VELS is almost as memory efficient as VE and considerably more memory efficient than CMOVE (Theorem 4.14).

5.7.2 EMPIRICAL COMPARISON

We use the Mining Day benchmark (Section 4.2.4) to empirically compare CMOVE and VELS. Mining Day is a well-structured MO-CoG problem with a variable number of agents, n, and a limited induced width.

We generated 30 Mining Day instances for increasing n and averaged the runtimes (Figure 5.5 (left)). At 160 agents, CMOVE has reached a runtime of $22s$. Exact VELS ($\varepsilon = 0$) can compute the complete CCS for a MO-CoG with 420 agents in the same time. This indicates that VELS greatly outperforms CMOVE on this structured 2-objective MO-CoG. Moreover, when we allow only 0.1% error ($\varepsilon = 0.001$), it takes only $1.1s$ to compute an ε-CCS for 420 agents, a speedup of over an order of magnitude.

To measure the additional speedups obtainable by further increasing ε, and to test VELS on very large problems, we generated Mining Day instances with $n \in \{250, 500, 1{,}000\}$. We averaged over 25 instances per value of ε. On these instances, exact VELS runs in $4.2s$ for $n = 250$, $30s$ for $n = 500$ and $218s$ for $n = 1{,}000$ on average. As expected, increasing ε leads to greater speedups (Figure 5.5 (right)). However, when ε is close to 0, i.e., the ε-CCS is close to the full CCS, the speedup is small. After ε has increased beyond a certain value (dependent on n), the decline

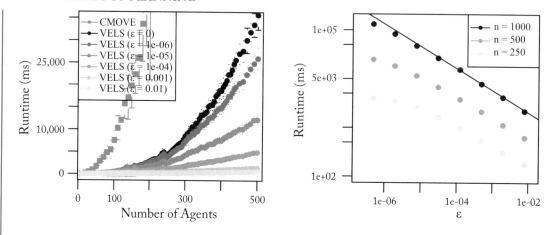

Figure 5.5: (Left) plot of the runtimes of CMOVE and VELS with different values of ε, for varying n (up to 500). (Right) log-logplot of the runtime of VELS on 250, 500, and 1,000 agent Mining Day instances, for varying values of ε.

becomes steady, shown as a line in the log-log plot. If ε increases by a factor 10, the runtime decreases by about a factor 1.6.

Thus, these results show that VELS can compute an exact CCS for much higher numbers of agents than CMOVE(1,000) in well-structured problems. In addition, they show that small values of ε enable large speedups, and that increasing ε leads to even bigger improvements in scalability.

5.8   OUTER LOOP METHODS FOR PCS PLANNING

At the beginning of this chapter, we mentioned that an outer loop approach does not work well for PCS planning. This is because the outer loop approach relies on solving a series of scalarized MODPs and, as discussed in Section 3.2.4, nonlinear f do not distribute over addition, it is unclear how to scalarize a problem with an additive reward structure (such as MOMDPs and MO-CoGs). Furthermore, OLS, the best performing outer loop approach, relies on the PWLC property of the scalarized value function, which holds only if f is linear.

Nonetheless, some attempts have been made to use an outer loop approach for PCS planning, by applying a scalarization to the (intermediate) value vectors, in a model-free method. For example, Van Moffaert et al. [2013] use the *Tchebycheff scalarization function* [Perny and Weng, 2010], which has the property that, for every policy in the Pareto front, there is a weight **w** for the Tchebycheff scalarization function for which that policy is optimal. Therefore, in principle, if we could optimally solve an SODP corresponding to a Tchebycheff-scalarized MODP, we could identify the entire Pareto front with a series of such SODPs. However, because Tchebycheff scalarization destroys the additivity of the reward structure, the guarantees of most standard methods for SODPs—which exploit additive reward structures—are nullified. Consequently, the method of Van Moffaert et al. [2013], which uses a single-objective subroutine that relies on the additivity of rewards via Bellman backups, does not have any guarantees with respect to optimality or convergence.

CHAPTER 6

Learning

In the previous chapters, we considered a planning setting in which the algorithm is given a model of the MODP and produces a coverage set. But what if the model is unknown at the time a coverage set needs to be produced? In that case, the algorithm must learn about the MODP from interaction with the environment. This is called the *reinforcement learning* or simply *learning* setting [Sutton and Barto, 1998, Wiering and Van Otterlo, 2012].

Multi-objective reinforcement learning (MORL) applies to all three of the use cases we described in the introduction (Figure 1.1): unknown weights, decision support, and known weights. However, in MORL, there is another important distinction, between *offline* and *online* learning.

In offline learning, the agent has an initial *learning phase* in which it interacts with the environment with the goal of finding a coverage set. The rewards it attains during learning do not matter. Next, it advances to the selection phase, which proceeds just as in the planning setting. Finally, in the execution phase, the selected policy is executed, at which point the rewards accrued *do* matter.

For example, imagine a luggage conveying system that has been installed at a new airport terminal. The system needs to balance maximizing throughput, minimizing energy usage, and minimizing the probability of damaging the system (i.e., minimizing expected maintenance cost). Since the system is too complex to manually build an accurate model, MORL is applied to learn the set of available possibly optimal policies through trial and error. Fortunately, the MORL algorithm is given three weeks before the terminal opens in which to learn, before actual travelers arrive. Thus, the rewards attained during learning thus do not matter, and this is an offline learning setting.

In online learning, learning and execution are intertwined. Therefore, the rewards received during learning also matter. For example, imagine a mining company that is optimizing its operations via reinforcement learning. The objectives correspond to different resources that can be sold on the open market. What the optimal policy is thus depends on the current prices on this market. It is infeasible to start learning again every time the prices change, so it is highly preferable to do MORL. Because the company's revenue depends on what it produces right now, the rewards received during learning are critically important. We first discuss the offline setting, and then the online setting.

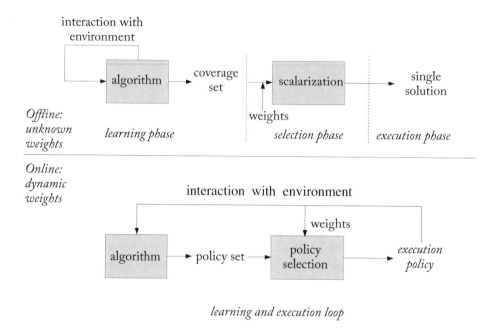

Figure 6.1: The difference between offline and online settings for multi-objective reinforcement learning: (top) the offline unknown weights scenario, (bottom) the online dynamic weights scenario.

6.1   OFFLINE MORL

In offline learning, the agent is free to experiment during the learning phase, and only the policy that is later executed matters. In the unknown weights scenario, this yields the setting depicted in the top part of Figure 6.1. The setup is identical to that of the unknown weights scenario for planning (Figure 1.1) except that the planning phase is replaced by a learning phase in which the agent learns from data gathered by interacting with the environment instead of planning via a model. The decision support and known weights scenarios (also shown in Figure 1.1) can be similarly modified.

In offline MORL, any type of learning algorithm can be employed. One approach is *model-based* learning, in which the data gathered while interacting with the environment is used to learn a model of the MODP, which can then be fed to a planner. The simplest approach is to gather all the data first, e.g., using a random policy. However, it is typically more effective to interleave planning and learning, i.e., to gather some data, plan on the learned model, and then use the resulting policy to gather more data, from which the model can be updated, etc. Wiering et al. [2014] proposed a model-based approach to learning a PCS of deterministic stationary policies, using *consistent multi-objective dynamic programming (CON-MODP)* [Wiering and De Jong, 2007] as the planning subroutine. CON-MODP is an inner loop method based on value iteration, similar to CHVI and PVI (Section 4.3.1).

However, model-based MORL remains an under-explored research topic, which is perhaps surprising, since it is a natural extension to single-objective model-based methods. Learning a model typically requires learning a *transition function*, which predicts how state evolves over time, and a *reward function*, which predicts rewards for each action in each state. The latter is more complicated in a multi-objective setting, since vector-valued rewards must be predicted instead of scalars. However, the former, which is typically the challenging part of model learning, is exactly the same as in the single-objective setting. Once the model is learned, any multi-objective planning method can be applied. Model-based approaches to MORL are thus a promising direction for future research.

MORL can also be performed without learning a model. Instead, the multi-objective value vectors in an (approximate) coverage set can be estimated directly from the data. For example, Moffaert and Nowé [2014] employ such an approach to learn an approximation of the PCS of deterministic, possibly non-stationary policies. To this end, they extend *Q-learning* [Watkins and Dayan, 1992], a popular single-objective model-free reinforcement learning method, to develop an inner loop approach.

MORL is also possible via heuristic policy search. For example, Handa [2009] applies a *multi-objective evolutionary algorithm* (MOEA) [Coello et al., 2007] to MORL. MOEAs keep a population of policies, introduce new policies into this population (typically based on mutations of and cross-overs between successful policies from the previous population), and remove the less successful policies. In this way, the population of policies gradually evolves into a better and better approximation of the coverage set.

6.2   ONLINE MORL

Unlike the offline setting, the online setting assumes that the rewards obtained during learning are important with respect to user utility. In other words, the policies that are learned are applied in practice during learning. We can model this setting as follows. Information about the scalarization function, f, and weights, $\mathbf{w}$ (corresponding to the user utility) is provided during learning; the learning algorithm produces a set of possible execution policies—which could be an intermediate estimate of the coverage set, but also a set of policies that takes factors like the necessary exploration for learning into account; a policy is selected from this set of policies using the latest available information about $\mathbf{w}$; the policy is executed, generating new interaction data, which, along with any new preference information, is fed back into the learning algorithm.

Figure 6.1 (bottom) shows an online variant of the unknown weights scenario, which we call the *dynamic weights scenario* [Natarajan and Tadepalli, 2005]. In this scenario, the weights $\mathbf{w}$ are determined by the environment, and are thus exactly known. However, $\mathbf{w}$ can change over time. For example, imagine that the objectives represent resources that can be bought and sold on an open market. In this case, $\mathbf{w}$ corresponds to the current prices of the resources on this market.

Natarajan and Tadepalli [2005] propose an algorithm called *dynamic multi-criteria reinforcement learning* that learns a CCS for a dynamic weights setting in which the policy values are the average reward per timestep. To learn the multi-objective value of a policy given a $\mathbf{w}$, they extend standard single-objective *H-learning* [Tadepalli and Ok, 1998], a model-based single-objective learning algorithm for average-reward RL. Instead of learning a scalar value for the problem scalarized by $\mathbf{w}$, as H-learning would do, their method keeps the maximizing value vector that underlies this scalarized value. When the weights change, the policy and value vectors for each state are stored in a set of previous policies. Then, for the new weight $\mathbf{w}'$, an initial policy is selected from the set of previous policies that maximize the scalarized value for $\mathbf{w}'$. This approach resembles the outer loop approach for planning (Chapter 5), in that single-objective methods are used for a given scalarization weight, $\mathbf{w}$. However, unlike in outer loop planning, these $\mathbf{w}$ are not selected by the algorithm but by an external stochastic process such as a market.

Besides the dynamic weights scenario, which corresponds to the unknown weights scenario, we could also have a *dynamic preferences scenario*, which would correspond to the decision support scenario. In this case, (partial) preference information about the current policy set would have to be provided by human decision makers. We are not aware of any work that investigates this in a *learning* scenario, though some work exists for the corresponding planning scenario [Benabbou and Perny, 2015, Wilson et al., 2015]. See Section 8.2.3 for a further discussion of including users in the planning or learning phases.

CHAPTER 7

Applications

In this chapter, we discuss a number of promising applications of multi-objective algorithms. We do not provide a comprehensive overview but instead describe a few selected examples that illustrate how multi-objective techniques can contribute to real-world problems. We divide the examples into three areas: (1) energy, (2) health, and (3) infrastructure and transportation. For each area, we present multiple examples that have recently been studied.

7.1 ENERGY

The production and consumption of energy is of vital importance in the modern world. Energy consumption incurs not only financial costs but environmental ones. Hence, in the Paris agreement of December 2015, 195 countries assert "the need for global emissions to peak as soon as possible . . . [and] to undertake rapid reductions thereafter in accordance with the best available science."[1] In the current economy, which depends heavily on fossil fuels, this implies that clean ways of producing energy must be found and/or energy must be conserved across society. Multi-objective decision making can help in this respect by enabling energy-consuming systems to be optimized, not only for traditional objectives such as financial cost, but also for the environmental damage they inflict or the non-renewable resources they consume.

For example, Tesauro et al. [2007] investigate maximizing the performance of web application servers, while minimizing power consumption, via multi-objective reinforcement learning. To this end, they use function approximation to learn the value of each objective separately. By doing so, they show that power consumption can be reduced by 10% while keeping the performance close to a target value.

Similarly, Kwak et al. [2012, 2014] investigate the rescheduling of meetings in an educational building at the University of Southern California, with a (bounded-parameter) multi-objective MDP. The agent uses planning to maximize the comfort levels of meeting participants while minimizing energy consumption. Subsequently, it suggests changes to the meeting schedule to the participants. In the most recent work [Kwak et al., 2014], the agent can re-plan if the participants decline the suggested change.

Multi-objective methods have also been used to optimize energy production. For example, in water reservoir management in Hanoi, a balance was sought between maximizing hydro-electric energy generation and minimizing flood risk [Castelletti et al., 2008, Pianosi et al., 2013]. This is a complex problem that is modeled as a continuous-state MOMDP, and has led to extensive

---

[1]http://ec.europa.eu/clima/policies/international/negotiations/paris/index_en.htm

work and several algorithms on this problem class, e.g., Castelletti et al. [2013, 2008], Giuliani et al. [2015, 2014], Pianosi et al. [2013] and Pirotta et al. [2015].

7.2   HEALTH

Health applications are naturally multi-objective. Sick people want to be cured, but they also want to minimize the risk of death, the severity of side effects, and financial cost.

For individual health and treatment, Lizotte et al. [2012] develop an agent that proposes multiple medical treatment plans for schizophrenia that offer different trade-offs between minimizing the severity of the symptoms, minimizing weight gain, and maximizing quality of life. Such decision support is essential when advising doctors and patients on treatment plans, as each patient has a different tolerance for side effects.

On a larger scale, i.e., in public health, multi-objective can also help. Soh and Demiris [2011] consider how to respond to possible anthrax attacks. When a letter containing white powder is opened, the Center for Disease Control may respond by, e.g., quarantine and further investigation through lab tests, and/or monitoring eventual symptoms. Thereby, they aim to minimize the loss of life, but also the probability of a (prolonged) false alarm, as this may put undue stress on those involved. Furthermore, investigation costs must also be moderated. One of the challenges in this case is the uncertainty about the true state of the incident. Therefore, Soh and Demiris [2011] propose a multi-objective partially observable MDP (MOPOMDP) model for which they optimize a finite state controller via evolutionary optimization.

7.3   INFRASTRUCTURE AND TRANSPORTATION

The infrastructure and transportation domains are also amenable to multi-objective methods. Because such systems affect society as a whole, there are often multiple stakeholders, whose interests may conflict. Multi-objective methods can assist in balancing these competing interests.

For example, imagine a community that must outsource maintenance of a traffic network to contractors. Such a *maintenance planning problem* (MPP) [Scharpff et al., 2013] has an enormous state space due to the multiple contractors, who each have maintenance tasks that can be incomplete, complete, or in progress. When the community wants to minimize both the traffic hindrance and the costs of maintenance, the MPP can be modeled with a finite-horizon multi-objective multi-agent MDP. Because the MPP exhibits special properties, Roijers et al. [2014a] solve the MPP using OLS (Section 5.3) and a problem-specific single-objective solver. The bottleneck in the MPP was the large state space, which prevented larger instances from being solved. However, after the state of the art for the single-objective MPP was improved by Scharpff et al. [2016], larger instances could be solved by the multi-objective method as well [Roijers, 2016]. Because it is relatively easy to exchange subroutines (but not to create inner loop methods from new single-objective methods), the outer loop approach was thus key to the creation of effective multi-objective planning methods for the MPP.

Another well-studied example in the literature is the control of traffic signal configurations at road-network junctions, e.g. [Kuyer et al., 2008, Wiering et al., 2004]. Khamis and Gomaa [2014] use multi-agent multi-objective reinforcement learning for traffic signal control. In contrast to earlier work, they define many objectives, e.g., minimizing trip waiting time, total trip time, the number of times a car has to stop, the risk of accidents, and junction waiting time, while maximizing flow rate, and satisfying green waves for platoons traveling on main roads. A single controller is learned, using a given scalarization, and compared to policies optimized for a single objective. Because of an overall improvement in optimization techniques, the learned policy achieves an improvement even in the individual objectives previous work had optimized with single-objective learning. More importantly, however, the performance in objectives that where not previously taken into account is now made explicit, leading to more insight into the effects of the policy. For example, the average number of stops per car decreases by about two thirds with respect to previous work. Because accelerating again after a stop costs a lot of fuel, decreasing the number of stops can significantly decrease environmental impact.

CHAPTER 8

Conclusions and Future Work

In this chapter, we present some concluding remarks and outline some opportunities for future work on multi-objective decision making.

8.1 CONCLUSIONS

The multi-objective methods presented in this book are useful, not only because there are many real-world problems that have multiple objectives, but because it is often impossible, infeasible, or undesirable to *a priori* scalarize such problems, i.e., to convert them to single-objective ones.

In order to make explicit under what circumstances special methods are needed to solve multi-objective problems, we identified three scenarios in which multi-objective methods are applicable: the unknown weights, decision support, and known weights scenarios. As well as providing motivation for the need for multi-objective methods, these scenarios also represent the three main ways these methods are applied in practice.

We also proposed a taxonomy that classifies multi-objective methods according to the applicable scenario, the scalarization function (which projects multi-objective values to scalar ones), and the type of policies that are allowed. We showed how these three aspects together determine the nature of an optimal solution, which can be a single policy, or a coverage set containing multiple policies.

Our taxonomy is based on a utility-based approach, which sees the scalarization function as part of the utility, and thus part of the problem definition. This contrasts with an *axiomatic approach*, which usually assumes the Pareto front is the appropriate solution. We showed that the utility-based approach can be used to justify the choice for a solution set. A surprising result of the taxonomy is that the Pareto front, by far the most widely used multi-objective solution concept, is only required in one special case: when the scalarization function might be nonlinear, and deterministic policies are required. In other cases where multiple policies are required, a convex coverage set, which is typically smaller and easier to compute, suffices.

We discussed the two main approaches to creating new algorithms for multi-objective decision making, by building on existing single-objective algorithms. In the inner loop approach, the maximizations and summations—which are necessary to compute the optimal policy and policy value of a single-objective decision problem—inside the single-objective algorithms are replaced with appropriate pruning and cross-sum operators between sets of value vectors.

In the outer loop approach, a series of scalarized problems are solved using the single-objective method as a subroutine in order to incrementally build up the coverage set for a multi-objective decision problem. Because these single-objective subroutines can be left unchanged, outer loop methods are generic frameworks: they can apply to any (scalarizable) multi-objective decision problem (such as MO-CoGs and MOMDPs). However, at present, only state-of-the-art outer loop planning methods that compute the CCS have guarantees with respect to convergence and quality of approximation.

We discussed optimistic linear support (OLS), a recent state-of-the-art outer loop method that exploits the piecewise linearity and convexity of the scalarized value function (under linear scalarization) in multi-objective decision problems. OLS solves a multi-objective decision problem as a series of scalarized problems, for different (linear) scalarization weights $\mathbf{w}$. At each iteration, OLS identifies the single weight vector $\mathbf{w}$ that can lead to the maximal possible improvement on a partial CCS it has already identified, and calls a single-objective subroutine to solve a scalarized instance of the problem for $\mathbf{w}$ (which is possible because of linear scalarization). By doing so, OLS is guaranteed to terminate in a finite number of iterations. Furthermore, when the single-objective subroutine is exact, i.e., guaranteed to find the optimal policy for the scalarized instance of the problem, OLS is guaranteed to return the exact CCS. Before termination, OLS can bound the quality loss if the partial CCS it has identified so far were used to approximate the exact CCS, in terms of lost scalarized value. In other words, it is an *anytime* algorithm, for which each intermediate CCS is an ε-CCS.

When a sufficient model of the environment cannot be provided to an agent beforehand, the agent learns about the environment through interaction. In multi-objective decision making, this leads to the multi-objective reinforcement learning (MORL) setting. We distinguish between offline and online MORL. In the former, there is a separate learning phase in which the agent gets to interact with and learn from the environment, while the rewards it attains during learning are not important. In this case, the same scenarios—unknown weights, decision support, and known weights—discussed for the planning setting apply. In the latter, the rewards attained during learning are important. In this case, the agent typically has (possibly limited) prior information about the preferences of the user, i.e., the scalarization function f and its weights $\mathbf{w}$. However, these preferences may change over time due to external factors, or be refined by the user(s). We discussed several algorithms for both the offline and the online MORL settings.

Finally, we discussed three different promising application domains for multi-objective methods. In the increasingly important energy domain, minimizing energy usage is explicitly balanced against other objectives, such as maximizing performance or financial gain. In the health domain, it is important to balance the impact for patients (e.g., via side-effects of a treatment or the psychological impact of quarantine) against public health or the probability of being cured. This is a naturally multi-objective setting in which obtaining a coverage set with different trade-offs can empower patients, doctors, and public health officials alike. Finally, in the infrastructure and transportation domain, the effects of, e.g., maintenance on the surrounding economy need to be balanced against the costs of such maintenance. Because traffic affects society as a whole, there are typically multiple stakeholders that have different preferences with respect to different objectives. In this case, having a coverage set can help when negotiating compromises between these different interests.

8.2   FUTURE WORK

In this section, we identify some opportunities for future work on multi-objective decision making.

8.2.1   SCALARIZATION OF EXPECTATION VS. EXPECTATION OF SCALARIZATION

In Section 1.1 in Definition 1.1, we defined the scalarized value of a policy, π, to be the result of applying the scalarization function f to the multi-objective value $\mathbf{V}^\pi$, i.e., $V_{\mathbf{w}}^\pi(s) = f(\mathbf{V}^\pi(s), \mathbf{w})$. Since $\mathbf{V}^\pi(s)$ is typically itself an expectation, this means that the scalarization function is applied *after* the expectation is computed, e.g., for MOMDPs (Definition 2.8):

$$V_{\mathbf{w}}^\pi(s) = f\left(\mathbf{V}^\pi(s), \mathbf{w}\right) = f\left(E\left[\sum_{k=0}^{\infty} \gamma^k \mathbf{r}_k \mid \pi, s_0 = s\right], \mathbf{w}\right).$$

This formulation, which we refer to as the *scalarization of the expected return* (SER), is standard in the literature. However, it is not the only conceivable option. It is also possible to define $V_{\mathbf{w}}^\pi(s)$ as the *expectation of the scalarized return* (ESR), e.g., for MOMDPs:

$$V_{\mathbf{w}}^\pi(s) = E\left[f\left(\sum_{k=0}^{\infty} \gamma^k \mathbf{r}_k, \mathbf{w}\right) \mid \pi, s_0 = s\right].$$

Which definition is used can critically affect which policies are preferred. For example, consider the following 2-objective infinite-horizon MOMDP, illustrated in Figure 8.1. There are four states (A, B, C, and D). The agent starts in state A and has two possible actions: a_1 transits to state B or C, each with probability of 0.5, and a_2 transits to state D with probability 1. Both actions lead to a $(0, 0)$ reward. In states B, C, and D there is only one action, which leads to a deterministic reward of $(3, 0)$ for B, $(0, 3)$ for C, and $(1, 1)$ for D.

The scalarization function just multiplies the two objectives together if the value in both objectives is greater than 0, and is 0, otherwise. Thus, for this MOMDP with only positive rewards, the scalarized value under SER is,

$$V_{\mathbf{w}}^\pi(s) = V_1^\pi(s) V_2^\pi(s),$$

and under ESR,

$$V_{\mathbf{w}}^\pi(s) = E\left[\left(\sum_{k=0}^{\infty} \gamma^k r_k^1\right)\left(\sum_{k=0}^{\infty} \gamma^k r_k^2\right) \mid \pi, s_0 = s\right],$$

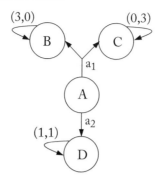

Figure 8.1: An MOMDP with two objectives and four states.

where r_k^i is the reward for the i-th objective on timestep k ($\mathbf{w}$ is not needed in this example since f involves no constants). If $\pi_1(A) = a_1$ and $\pi_2(A) = a_2$, then the multi-objective values are $\mathbf{V}^{\pi_1}(A) = (1.5\gamma/(1-\gamma), 1.5\gamma/(1-\gamma))$ and $\mathbf{V}^{\pi_2}(A) = (\gamma/(1-\gamma), \gamma/(1-\gamma))$.

Under SER, this leads to scalarized values of $V^{\pi_1}(A) = (1.5\gamma/(1-\gamma))^2$ and $V^{\pi_2}(A) = (\gamma/(1-\gamma))^2$ and consequently π_1 is preferred. Under ESR, however, we have $V^{\pi_1}(A) = 0$ and $V^{\pi_2}(A) = (\gamma/(1-\gamma))^2$ and thus π_2 is preferred.

Intuitively, the SER formulation is appropriate when the policy will be used many times and *return accumulates across episodes*, e.g., because the same user is using the policy each time. Then, scalarizing the expected reward makes sense and π_1 is preferable because in expectation it will accumulate more return in both objectives.

However, if the policy will only be used a few times or the return does not accumulate across episodes, e.g., because each episode is conducted for a different user, then the ESR formulation seems more appropriate. In this case, the expected return before scalarization is not of interest and π_2 is preferable because π_1 will always yield zero scalarized return on any given episode.

This issue is related to the question of whether or not to allow stochastic policies, which was discussed in Section 3.1.3. When a policy will be used many times and return accumulates across episodes, then it is logical to allow stochastic policies in order to increase the scalarized expected return. When a policy is only be used a few times or the return does not accumulate across episodes, it is inappropriate to allow stochastic policies.

To our knowledge, there is no literature on multi-objective decision-theoretic planning and learning that employs the ESR formulation, even though there are many real-world scenarios in which it seems more appropriate. For example, in the medical application of Lizotte et al. [2010], each patient gets only one episode to treat his or her illness, and needs to balance the effectiveness of the treatment against the severity of the side effects. Therefore, the patient is clearly interested in maximizing ESR, not SER. Thus, we believe that developing methods for the ESR formulation is a critical direction for future research in multi-objective decision making.

8.2.2   OTHER DECISION PROBLEMS

For the sake of brevity, this book focuses on multi-objective versions of coordination graphs and Markov decision processes. However, there are many other single-objective decision problems for which multi-objective variants have been considered, such as the *multi-agent Markov decision process* (MMDP) [Scharpff et al., 2015] and the *partially observable Markov decision process* [Roijers et al., 2015c]. There are also several single-objective decision problems for which, to our knowledge, no multi-objective treatment exists.

For example, *collaborative Bayesian games* (CBGs) [Oliehoek et al., 2012] are single-shot, multi-agent models like CoGs. Unlike CoGs, however, each agent receives a private observation, called a type θ, before it needs to act. The distributions over types for each agent are common knowledge. Deterministic policies in a CBG specify an action for each agent for every type of that agent. CBGs can be flattened to a large CoG by enumerating the possible mappings from types to actions for each agent and seeing these mappings as that agent's local action space, A_i. This same reduction could be made for a multi-objective version of a CBG to a MO-CoG.

CBGs are often used as a subproblem [MacDermed and Isbell, 2013] in the sequential version of the problem, the *decentralized partially observable Markov decision process* (Dec-POMDP) [Bernstein et al., 2002, Oliehoek, 2010], for which, to our knowledge, no multi-objective extension has been considered. Recently, major strides in solving single-objective Dec-POMDPs have been made. In particular, it has been shown that there is a reduction from Dec-POMDP to a special type of centralized POMDP called a *non-observable Markov decision process* (NOMDP) [Dibangoye et al., 2013, MacDermed and Isbell, 2013, Nayyar et al., 2013, Oliehoek and Amato, 2014]. This allows POMDP solution methods to be employed in the context of Dec-POMDPs. The reduction to NOMDPs is potentially useful for solving multi-objective Dec-POMDPs as well. Specifically, an equivalent reduction from an MO-Dec-POMDP to a *multi-objective NOMDP* (MONOMDP) could be made, for which an existing multi-objective POMDP algorithm [Roijers et al., 2015c] would apply.

8.2.3   USERS IN THE LOOP

In this book, we considered scenarios in which the human user is either not directly involved (the unknown and known weights scenarios) or is involved only in a separate selection phase that occurs after the coverage set has been completely computed (the decision support scenario).

In the latter scenario, waiting for the selection phase to involve the user may be suboptimal, for two reasons. First, the coverage set might be too large for the user to analyze, and second, computing the coverage set may involve a lot of wasted effort to identify policies that could have been ruled out if more was known about the user's preferences.

An interesting way to mitigate both issues is to involve the decision maker while planning. In particular, when we produce intermediate results during the planning phase, we might elicit more preference information from the decision maker by asking questions about these intermediate results [Benabbou and Perny, 2015, Wilson et al., 2015]. We believe that such an approach would be particularly compatible with OLS, as decision makers can be presented with values of complete policies from the partial CCS between iterations of OLS. The answers that the user provides can be used by OLS to quickly reduce the interesting region of the weight space, leading to large speedups.

Bibliography

Arnborg, S. (1985). Efficient algorithms for combinatorial problems on graphs with bounded decomposability—a survey. *BIT Numerical Mathematics*, 25(1):1–23, 1985. DOI: 10.1007/bf01934985. 44

Avis, D. and Devroye, L. (2000). Estimating the number of vertices of a polyhedron. *Information Processing Letters*, 73(3):137–143, 2000. DOI: 10.1016/s0020-0190(00)00011-9. 76

Barrett, L. and Narayanan, S. (2008). Learning all optimal policies with multiple criteria. In *ICML: Proc. of the 25th International Conference on Machine Learning*, pages 41–47, 2008. DOI: 10.1145/1390156.1390162. 20, 59

Bellman, R. (1957a). *Dynamic Programming*. Princeton University Press, Princeton, NJ, 1957a. DOI: 10.1126/science.153.3731.34. 15, 58

Bellman, R. E. (1957b). A Markovian decision process. *Journal of Mathematics and Mechanics*, 6:679–684, 1957b. DOI: 10.1512/iumj.1957.6.56038. 14

Benabbou, N. and Perny, P. (2015). Incremental weight elicitation for multiobjective state space search. In *Proc. of the 29th AAAI Conference on Artificial Intelligence*, pages 1093–1099, 2015. 90, 99

Bernstein, D. S., Givan, R., Immerman, N., and Zilberstein, S. (2002). The complexity of decentralized control of Markov decision processes. *Mathematics of Operations Research*, 27(4):819–840, 2002. DOI: 10.1287/moor.27.4.819.297. 99

Bertsimas, D. and Tsitsiklis, J. (1997). *Introduction to Linear Optimization*. Athena Scientific, Nashua (NH), 1997. 71

Bishop, C. M. (2006). *Pattern Recognition and Machine Learning*. Springer, New York, 2006. 11

Boutilier, C. (1996). Planning, learning and coordination in multiagent decision processes. In *TARK: Proc. of the 6th Conference on Theoretical Aspects of Rationality and Knowledge*, pages 195–210, 1996. 10

Boutilier, C., Dean, T., and Hanks, S. (1999). Decision-theoretic planning: Structural assumptions and computational leverage. *Journal of Artificial Intelligence Research*, 11:1–94, 1999. 15, 16

Bryce, D. (2008). The value(s) of probabilistic plans. In *Workshop on a Reality Check for Planning and Scheduling under Uncertainty, ICAPS-08*, 2008. 2

Bryce, D., Cushing, W., and Kambhampati, S. (2007). Probabilistic planning is multi-objective! Technical Report 08-006, Arizona State University, 2007. 2

Busoniu, L., Babuska, R., and De Schutter, B. (2008). A comprehensive survey of multiagent reinforcement learning. *IEEE Transactions on Systems, Man, and Cybernetics*, 38(2):156–172, 2008. DOI: 10.1109/tsmcc.2007.913919. 1

Cassandra, A., Littman, M., and Zhang, N. (1997). Incremental pruning: A simple, fast, exact method for partially observable Markov decision processes. In *UAI: Proc. of the 13th Conference on Uncertainty in Artificial Intelligence*, pages 54–61, 1997. 51

Castelletti, A., Pianosi, F., and Restelli, M. (2013). A multiobjective reinforcement learning approach to water resources systems operation: Pareto frontier approximation in a single run. *Water Resources Research*, 2013. DOI: 10.1002/wrcr.20295. 17, 92

Castelletti, A., Pianosi, F., and Soncini-Sessa, R. (2008). Water reservoir control under economic, social and environmental constraints. *Automatica*, 44:1595–1607, 2008. DOI: 10.1016/j.automatica.2008.03.003. 17, 91, 92

Chalkiadakis, G., Elkind, E., and Wooldridge, M. (2011). Computational aspects of cooperative game theory. *Synthesis Lectures on Artificial Intelligence and Machine Learning*, 5(6):1–168, 2011. DOI: 10.2200/s00355ed1v01y201107aim016. 10

Cheng, H.-T. (1988). *Algorithms for Partially Observable Markov Decision Processes*. Ph.D. thesis, University of British Columbia, Vancouver. 67, 68, 73

Chu, W. and Ghahramani, Z. (2005). Preference learning with Gaussian processes. In *ICML: Proc. of the 22nd International Conference on Machine Learning*, pages 137–144, 2005. DOI: 10.1145/1102351.1102369. 4

Clemen, R. T. (1997). *Making Hard Decisions: An Introduction to Decision Analysis*, 2nd ed. South-Western College Pub, 1997. 5

Clempner, J. B. (2016). Necessary and sufficient Karush-Kuhn-Tucker conditions for multiobjective Markov chains optimality. *Automatica*, 71:135–142, 2016. DOI: 10.1016/j.automatica.2016.04.044. 28

Coello, C. C., Lamont, G. B., and Van Veldhuizen, D. A. (2007). *Evolutionary algorithms for solving multi-objective problems*. Springer Science & Business Media, 2007. DOI: 10.1007/978-1-4757-5184-0. 89

Dechter, R. (1998). Bucket elimination: A unifying framework for probabilistic inference. In *Learning in Graphical Models*, pages 75–104. Springer, Netherlands, 1998. DOI: 10.1007/978-94-011-5014-9_4. 43, 44, 46

Dibangoye, J. S., Amato, C., Buffet, O., and Charpillet, F. (2013). Optimally solving Dec-POMDPs as continuous-state MDPs. In *IJCAI: Proc. of the 22nd International Joint Conference on Artificial Intelligence*, 2013. 99

Feng, Z. and Zilberstein, S. (2004). Region-based incremental pruning for POMDPs. In *UAI: Proc. of the 20th Conference on Uncertainty in Artificial Intelligence*, pages 146–153, 2004. 41

Franklin, S. and Graesser, A. (1997). Is it an agent, or just a program?: A taxonomy for autonomous agents. In *Intelligent Agents III Agent Theories, Architectures, and Languages*, pages 21–35, Springer, 1997. DOI: 10.1007/bfb0013570. 1

Giuliani, M., Castelletti, A., Pianosi, F., Mason, E., and Reed, P. M. (2015). Curses, trade-offs, and scalable management: Advancing evolutionary multiobjective direct policy search to improve water reservoir operations. *Journal of Water Resources Planning and Management*, page 04015050, 2015. DOI: 10.1061/(asce)wr.1943-5452.0000570. 17, 92

Giuliani, M., Galelli, S., and Soncini-Sessa, R. (2014). A dimensionality reduction approach for many-objective Markov decision processes: Application to a water reservoir operation problem. *Environmental Modelling and Software*, 57:101–114, 2014. DOI: 10.1016/j.envsoft.2014.02.011. 92

Graham, R. L. (1972). An efficient algorithm for determining the convex hull of a finite planar set. *Information Processing Letters*, 1(4):132–133, 1972. DOI: 10.1016/0020-0190(72)90045-2. 41

Guestrin, C., Koller, D., and Parr, R. (2002). Multiagent planning with factored MDPs. In *NIPS: Advances in Neural Information Processing Systems 15*, pages 1523–1530, 2002. 11, 43, 46

Guizzo, E. (2011). How Google's self-driving car works. *IEEE Spectrum Online*, 2011. 1

Handa, H. (2009). Solving multi-objective reinforcement learning problems by EDA-RL—acquisition of various strategies. In *ISDA: Proc. of the 9th International Conference on Intelligent Systems Design and Applications*, pages 426–431, 2009. DOI: 10.1109/isda.2009.92. 89

Henk, M., Richter-Gebert, J., and Ziegler, G. M. (1997). Basic properties of convex polytopes. In *Handbook of Discrete and Computational Geometry*, Ch. 13, pages 243–270. CRC Press, Boca Raton, FL, 1997. DOI: 10.1201/9781420035315.pt2. 76

Howard, R. A. (1960). *Dynamic Programming and Markov Decision Processes*. MIT Press, 1960. 16

Igarashi, A. and Roijers, D. M. (2017). Multi-criteria coalition formation games. In *Proc. of the 8th Workshop on Cooperative Games in Multiagent Systems*, 2017. 10

Jarvis, R. A. (1973). On the identification of the convex hull of a finite set of points in the plane. *Information Processing Letters*, 2(1):18–21, 1973. DOI: 10.1016/0020-0190(73)90020-3. 25

Kaelbling, L., Littman, M., and Cassandra, A. (1998). Planning and acting in partially observable stochastic domains. *Artificial Intelligence*, 101:99–134, 1998. DOI: 10.1016/s0004-3702(98)00023-x. 25, 67

Kaibel, V. and Pfetsch, M. E. (2003). Some algorithmic problems in polytope theory. In *Algebra, Geometry and Software Systems*, pages 23–47. Springer, 2003. DOI: 10.1007/978-3-662-05148-1_2. 76

Ketter, W., Peters, M., and Collins, J. (2013). Autonomous agents in future energy markets: The 2012 power trading agent competition. In *BNAIC: Proc. of the 25th Benelux Conference on Artificial Intelligence*, pages 328–329, 2013. 1

Khamis, M. A. and Gomaa, W. (2014). Adaptive multi-objective reinforcement learning with hybrid exploration for traffic signal control based on cooperative multi-agent framework. *Engineering Applications of Artificial Intelligence*, 29:134–151, 2014. DOI: 10.1016/j.engappai.2014.01.007. 93

Kim, I. Y. and de Weck, O. L. (2005). Adaptive weighted-sum method for bi-objective optimization: Pareto front generation. *Structural and Multidisciplinary Optimization*, 29(2):149–158, 2005. DOI: 10.1007/s00158-004-0465-1. 65

Kok, J. and Vlassis, N. (2006). Collaborative multiagent reinforcement learning by payoff propagation. *Journal of Machine Learning Research*, 7:1789–1828, 2006. 43

Kok, J. R. and Vlassis, N. (2004). Sparse cooperative Q-learning. In *ICML: Proc. of the 21st International Conference on Machine Learning*, pages 61–68, 2004. DOI: 10.1145/1015330.1015410. 11

Koller, D. and Friedman, N. (2009). *Probabilistic Graphical Models: Principles and Techniques*. MIT Press, 2009. 44, 51

Kuyer, L., Whiteson, S., Bakker, B., and Vlassis, N. (2008). Multiagent reinforcement learning for urban traffic control using coordination graphs. In *Joint European Conference on Machine Learning and Knowledge Discovery in Databases*, pages 656–671, 2008. DOI: 10.1007/978-3-540-87479-9_61. 93

Kwak, J., Varakantham, P., Maheswarn, R., Tambe, M., Jazizadeh, F., Kavulya, G., Klein, L., Becerik-Gerber, B., Hayes, T., and Wood, W. (2012). SAVES: A sustainable multiagent application to conserve building energy considering occupants. In *11th International Conference on Autonomous Agents and Multiagent Systems*, pages 21–28, 2012. 17, 91

Kwak, J.-y., Kar, D., Haskell, W. B., Varakantham, P., and Tambe, M. (2014). Building THINC: User incentivization and meeting rescheduling for energy savings. In *Proc. of the International Conference on Autonomous Agents and Multi-agent Systems*, pages 925–932. International Foundation for Autonomous Agents and Multiagent Systems, 2014. 91

Lizotte, D., Bowling, M., and Murphy, S. (2010). Efficient reinforcement learning with multiple reward functions for randomized clinical trial analysis. In *ICML: Proc. of the 27th International Conference on Machine Learning*, pages 695–702, 2010. 17, 21, 98

Lizotte, D. J., Bowling, M., and Murphy, S. A. (2012). Linear fitted-q iteration with multiple reward functions. *Journal of Machine Learning Research*, 13:3253–3295, 2012. 17, 92

MacDermed, L. C. and Isbell, C. (2013). Point based value iteration with optimal belief compression for Dec-POMDPs. In *NIPS: Advances in Neural Information Processing Systems 26*, pages 100–108, 2013. 99

Marinescu, R. (2011). Efficient approximation algorithms for multi-objective constraint optimization. In *ADT: Proc. of the 2nd International Conference on Algorithmic Decision Theory*, pages 150–164, 2011. DOI: 10.1007/978-3-642-24873-3_12. 10, 11

Mateescu, R. and Dechter, R. (2005). The relationship between AND/OR search and variable elimination. In *UAI: Proc. of the 21st Conference on Uncertainty in Artificial Intelligence*, pages 380–387, 2005. 46

McMullen, P. (1970). The maximum numbers of faces of a convex polytope. *Mathematika*, 17(2):179–184, 1970. DOI: 10.1112/s0025579300002850. 76

Moffaert, K. V. and Nowé, A. (2014). Multi-objective reinforcement learning using sets of Pareto dominating policies. *Journal of Machine Learning Research*, 15:3483–3512, 2014. 89

Monostori, L., Váncza, J., and Kumara, S. R. (2006). Agent-based systems for manufacturing. *CIRP Annals-Manufacturing Technology*, 55(2):697–720, 2006. DOI: 10.1016/j.cirp.2006.10.004. 1

Natarajan, S. and Tadepalli, P. (2005). Dynamic preferences in multi-criteria reinforcement learning. In *ICML*, 2005. DOI: 10.1145/1102351.1102427. 20, 89

Nayyar, A., Mahajan, A., and Teneketzis, D. (2013). Decentralized stochastic control with partial history sharing: A common information approach. *IEEE Transactions on Automatic Control*, 58:1644–1658, 2013. DOI: 10.1109/tac.2013.2239000. 99

Oliehoek, F. A. (2010). *Value-based Planning for Teams of Agents in Stochastic Partially Observable Environments*. Ph.D. thesis, University of Amsterdam, 2010. DOI: 10.5117/9789056296100. 1, 99

Oliehoek, F. A. and Amato, C. (2014). Dec-POMDPs as non-observable MDPs. IAS technical report IAS-UVA-14-01, Amsterdam, The Netherlands, 2014. 99

Oliehoek, F. A. and Amato, C. (2016). *A Concise Introduction to Decentralized POMDPs*. Springer Briefs in Intelligent Systems. Springer, 2016. Authors' pre-print. DOI: 10.1007/978-3-319-28929-8. 10

Oliehoek, F. A., Spaan, M. T. J., and Witwicki, S. (2015). Factored upper bounds for multiagent planning problems under uncertainty with non-factored value functions. pages 1645–1651, 2015. 7

Oliehoek, F. A., Whiteson, S., and Spaan, M. T. J. (2012). Exploiting structure in cooperative Bayesian games. In *UAI: Proc. of the 28th Conference on Uncertainty in Artificial Intelligence*, pages 654–664, 2012. 99

Ong, S. C. W., Png, S. W., Hsu, D., and Lee, W. S. (2010). Planning under uncertainty for robotic tasks with mixed observability. *The International Journal of Robotics Research*, 29(8):1053–1068, 2010. DOI: 10.1177/0278364910369861. 5

Pardoe, D. M. (2011). *Adaptive Trading Agent Strategies Using Market Experience*. Ph.D. thesis, University of Texas at Austin, 2011. 1

Pareto, V. (1896). *Manuel d'économie Politique*, 1896. DOI: 10.3917/droz.paret.1981.02. 21

Perny, P. and Weng, P. (2010). On finding compromise solutions in multiobjective Markov decision processes. In *ECAI Multidisciplinary Workshop on Advances in Preference Handling*, pages 55–60, 2010. DOI: 10.3233/978-1-60750-606-5-969. 31, 84

Perny, P., Weng, P., Goldsmith, J., and Hanna, J. P. (2013). Approximation of Lorenz-optimal solutions in multiobjective Markov decision processes. In *UAI: Proc. of the 29th Conference on Uncertainty In Artificial Intelligence*, pages 508–517, 2013. 35

Pianosi, F., Castelletti, A., and Restelli, M. (2013). Tree-based fitted q-iteration for multi-objective Markov decision processes in water resource management. *Journal of Hydroinformatics*, 15(2):258–270, 2013. DOI: 10.2166/hydro.2013.169. 91, 92

Pirotta, M., Parisi, S., and Restelli, M. (2015). Multi-objective reinforcement learning with continuous Pareto frontier approximation. In *29th AAAI Conference on Artificial Intelligence*, pages 2928–2934, 2015. 92

Roijers, D. M. (2016). *Multi-Objective Decision-theoretic Planning*. Ph.D. thesis, University of Amsterdam, 2016. DOI: 10.1145/3008665.3008670. xiii, 46, 63, 67, 76, 80, 92

Roijers, D. M., Scharpff, J., Spaan, M. T., Oliehoek, F. A., de Weerdt, M., and Whiteson, S. (2014a). Bounded approximations for linear multi-objective planning under uncertainty. In *ICAPS: Proc. of the 24th International Conference on Automated Planning and Scheduling*, pages 262–270, 2014a. 76, 92

Roijers, D. M., Vamplew, P., Whiteson, S., and Dazeley, R. (2013a). A survey of multi-objective sequential decision-making. *Journal of Artificial Intelligence Research*, 47:67–113, 2013a. DOI: 10.1613/jair.3987. xiii, 2, 16

Roijers, D. M., Whiteson, S., Ihler, A. T., and Oliehoek, F. A. (2015a). Variational multi-objective coordination. In *MALIC: NIPS Workshop on Learning, Inference and Control of Multi-Agent Systems*, 2015a. 79, 80

Roijers, D. M., Whiteson, S., and Oliehoek, F. (2013b). Computing convex coverage sets for multi-objective coordination graphs. In *ADT: Proc. of the 3rd International Conference on Algorithmic Decision Theory*, pages 309–323, 2013b. DOI: 10.1007/978-3-642-41575-3_24. 43, 47, 50

Roijers, D. M., Whiteson, S., and Oliehoek, F. A. (2013c). Multi-objective variable elimination for collaborative graphical games. In *AAMAS: Proc. of the 12th International Joint Conference on Autonomous Agents and Multiagent Systems*, pages 1209–1210, 2013c. Extended Abstract. 10

Roijers, D. M., Whiteson, S., and Oliehoek, F. A. (2014b). Linear support for multi-objective coordination graphs. In *AAMAS: Proc. of the 13th International Joint Conference on Autonomous Agents and Multi-Agent Systems*, pages 1297–1304, 2014b. 63, 67, 82

Roijers, D. M., Whiteson, S., and Oliehoek, F. A. (2015b). Computing convex coverage sets for faster multi-objective coordination. *Journal of Artificial Intelligence Research*, 52:399–443, 2015b. DOI: 10.1007/978-3-642-41575-3_24. 12, 43, 47, 50, 51, 56, 58, 63, 67, 82

Roijers, D. M., Whiteson, S., and Oliehoek, F. A. (2015c). Point-based planning for multi-objective POMDPs. In *IJCAI: Proc. of the 24th International Joint Conference on Artificial Intelligence*, pages 1666–1672, 2015c. 80, 99

Rollón, E. (2008). *Multi-objective Optimization for Graphical Models*. Ph.D. thesis, Universitat Politècnica de Catalunya, Barcelona, 2008. 10, 43

Rollón, E. and Larrosa, J. (2006). Bucket elimination for multiobjective optimization problems. *Journal of Heuristics*, 12:307–328, 2006. DOI: 10.1007/s10732-006-6726-y. 43, 47, 50, 53

Rosenthal, A. (1977). Nonserial dynamic programming is optimal. In *Proc. of the 9th Annual ACM Symposium on Theory of Computing*, pages 98–105, 1977. DOI: 10.1145/800105.803399. 43, 46

Russell, S. and Norvig, P. (1995). *Artificial Intelligence: A Modern Approach*. Prentice-Hall, Englewood Cliffs, 1995. 1

Scharpff, J., Roijers, D. M., Oliehoek, F. A., Spaan, M. T., and de Weerdt, M. M. (2015). Solving multi-agent MDPs optimally with conditional return graphs. In *MSDM: Proc. of the AAMAS Workshop on Multi-Agent Sequential Decision Making in Uncertain Domains*, 2015. 99

Scharpff, J., Roijers, D. M., Oliehoek, F. A., Spaan, M. T., and de Weerdt, M. M. (2016). Solving transition-independent multi-agent MDPs with sparse interactions. In *Proc. of the 30th AAAI Conference on Artificial Intelligence*. To Appear. 10, 92

Scharpff, J., Spaan, M. T. J., de Weerdt, M. M., and Volker, L. (2013). Planning under uncertainty for coordinating infrastructural maintenance. In *Proc. of the International Conference on Automated Planning and Scheduling*, pages 425–433, 2013. 92

Sen, A. K. (1995). *Collective Choice and Social Welfare*. Elsevier, Amsterdam, 1995. 35

Shortreed, S. M., Laber, E., Lizotte, D. J., Stroup, T. S., Pineau, J., and Murphy, S. A. (2011). Informing sequential clinical decision-making through reinforcement learning: An empirical study. *Machine Learning*, 84(1-2):109–136, 2011. DOI: 10.1007/s10994-010-5229-0. 21

Soh, H. and Demiris, Y. (2011). Multi-reward policies for medical applications: Anthrax attacks and smart wheelchairs. In *GECCO: Proc. of the 13th Annual Conference on Genetic and Evolutionary Computation*, pages 471–478, 2011. DOI: 10.1145/2001858.2002036. 92

Sutton, R. S. and Barto, A. G. (1998). *Introduction to Reinforcement Learning*. MIT Press, 1998. 15, 58, 87

Tadepalli, P. and Ok, D. (1998). Model-based average reward reinforcement learning. *AI Journal*, (100):177–223, 1998. DOI: 10.1016/s0004-3702(98)00002-2. 90

Tambe, M. (2011). *Security and Game Theory: Algorithms, Deployed Systems, Lessons Learned*. Cambridge University Press, 2011. DOI: 10.1017/cbo9780511973031. 1

Tesauro, G., Das, R., Chan, H., Kephart, J. O., Lefurgy, C., Levine, D. W., and Rawson, F. (2007). Managing power consumption and performance of computing systems using reinforcement learning. In *NIPS: Advances in Neural Information Processing Systems 20*, pages 1497–1504, 2007. 1, 91

Thiébaux, S., Gretton, C., Slaney, J. K., Price, D., Kabanza, F., et al. (2006). Decision-theoretic planning with non-markovian rewards. *Journal of Artificial Intelligence Research*, 25:17–74, 2006. 1

Vamplew, P., Dazeley, R., Barker, E., and Kelarev, A. (2009). Constructing stochastic mixture policies for episodic multiobjective reinforcement learning tasks. In *AI: Proc. of the 22nd Australasian Joint Conference on Artificial Intelligence*, pages 340–349, 2009. DOI: 10.1007/978-3-642-10439-8_35. 28, 31

Vamplew, P., Dazeley, R., Berry, A., Dekker, E., and Issabekov, R. (2011). Empirical evaluation methods for multiobjective reinforcement learning algorithms. *Machine Learning*, 84(1-2):51–80, 2011. DOI: 10.1007/s10994-010-5232-5. 19

Van Doorn, J., Odijk, D., Roijers, D. M., and de Rijke, M. (2016). Balancing relevance criteria through multi-objective optimization. In *39th International ACM SIGIR Conference on Research and Development in Information Retrieval*, 2016. DOI: 10.1145/2911451.2914708. 2, 80

Van Moergestel, L. J. (2014). *Agent Technology in Agile Multiparallel Manufacturing and Product Support*. Ph.D. thesis, Utrecht University, 2014. DOI: 10.13140/RG.2.1.3113.0487. 1

Van Moffaert, K. (2016). *Multi-criteria Reinforcement Learning for Sequential Decision Making Problems*. Ph.D. thesis, Vrije Universiteit Brussel, Brussels, Belgium, 2016. 59

Van Moffaert, K., Brys, T., Chandra, A., Esterle, L., Lewis, P. R., and Nowé, A. (2014). A novel adaptive weight selection algorithm for multi-objective multi-agent reinforcement learning. In *IJCNN: Proc. of the International Joint Conference on Neural Networks*, pages 2306–2314, 2014. DOI: 10.1109/ijcnn.2014.6889637. 65

Van Moffaert, K., Drugan, M. M., and Nowé, A. (2013). Scalarized multi-objective reinforcement learning: Novel design techniques. In *IEEE Symposium on Adaptive Dynamic Programming and Reinforcement Learning (ADPRL)*, pages 191–199, 2013. DOI: 10.1109/adprl.2013.6615007. 84, 85

Wakuta, K. (1999). A note on the structure of value spaces in vector-valued Markov decision processes. *Mathematical Methods of Operations Research*, 49(1):77–85, 1999. DOI: 10.1007/PL00020907. 32

Wakuta, K. and Togawa, K. (1998). Solution procedures for Markov decision processes. *Optimization: A Journal of Mathematical Programming and Operations Research*, 43(1):29–46, 1998. DOI: 10.1080/02331939808844372. 32

Watkins, C. J. and Dayan, P. (1992). Q-learning. *Machine Learning*, 8(3-4):279–292, 1992. DOI: 10.1007/bf00992698. 89

White, C. C. and Kim, K. M. (1980). Solution procedures for solving vector criterion Markov decision processes. *Large Scale Systems*, 1:129–140, 1980. 67

White, D. (1982). Multi-objective infinite-horizon discounted Markov decision processes. *Journal of Mathematical Analysis and Applications*, 89(2):639–647, 1982. DOI: 10.1016/0022-247x(82)90122-6. 27, 28, 59

Wiering, M., Vreeken, J., Van Veenen, J., and Koopman, A. (2004). Simulation and optimization of traffic in a city. In *IEEE Intelligent Vehicles Symposium*, pages 453–458, 2004. DOI: 10.1109/ivs.2004.1336426. 93

Wiering, M. A. and De Jong, E. D. (2007). Computing optimal stationary policies for multi-objective Markov decision processes. In *IEEE International Symposium on Approximate Dynamic Programming and Reinforcement Learning*, pages 158–165, 2007. DOI: 10.1109/adprl.2007.368183. 88

Wiering, M. A. and Van Otterlo, M. (2012). Reinforcement learning: State-of-the-art. In *Adaptation, Learning, and Optimization*, volume 12. Springer, 2012. DOI: 10.1007/978-3-642-27645-3. 1, 58, 87

Wiering, M. A., Withagen, M., and Drugan, M. M. (2014). Model-based multi-objective reinforcement learning. In *ADPRL: Proc. of the IEEE Symposium on Adaptive Dynamic Programming and Reinforcement Learning*, pages 1–6, 2014. DOI: 10.1109/adprl.2014.7010622. 88

Wilson, N., Razak, A., and Marinescu, R. (2015). Computing possibly optimal solutions for multi-objective constraint optimisation with tradeoffs. In *IJCAI: Proc. of the 24th International Joint Conference on Artificial Intelligence*, pages 815–821, 2015. 90, 99

Zintgraf, L. M., Kanters, T. V., Roijers, D. M., Oliehoek, F. A., and Beau, P. (2015). Quality assessment of MORL algorithms: A utility-based approach. In *Benelearn: Proc. of the 24th Belgian-Dutch Conference on Machine Learning*, 2015. 7, 34

Zitzler, E., Thiele, L., Laumanns, M., Fonseca, C. M., and Da Fonseca, V. G. (2003). Performance assessment of multiobjective optimizers: An analysis and review. *IEEE Transactions on Evolutionary Computation*, 7(2):117–132, 2003. DOI: 10.1109/tevc.2003.810758. 34

Authors' Biographies

DIEDERIK M. ROIJERS

Diederik M. Roijers completed his master's in Computing Science at Utrecht University before obtaining his Ph.D. in Artificial Intelligence under the supervision of Shimon Whiteson and Frans A. Oliehoek at the University of Amsterdam in 2016. He then joined the University of Oxford as a postdoctoral research assistant. He was awarded a Postdoctoral Fellowship Grant from the FWO (Research Foundation – Flanders) and started as an FWO Postdoctoral Fellow at the Vrije Universiteit Brussel in October 2016.

His research focuses on creating intelligent autonomous systems that assist humans in solving complex problems, especially those with multiple objectives. To this end, he focuses on decision-theoretic planning and learning, which enable agents to use mathematical models to reason about the environments in which they operate. In the multi-objective problems he has been studying, the agents produce a set of possibly optimal policies that offer different trade-offs with respect to the objectives, to help users make an informed decision.

SHIMON WHITESON

Shimon Whiteson studied English and Computer Science at Rice University before completing his doctorate in Computer Science under the supervision of Peter Stone at the University of Texas at Austin in 2007. He then spent eight years as an Assistant and then an Associate Professor at the University of Amsterdam before joining the University of Oxford as an Associate Professor in 2015. He was awarded an ERC Starting Grant from the European Research Council in 2014.

His research focuses on artificial intelligence with the goal of designing, analyzing, and evaluating the algorithms that enable computational systems to acquire and execute intelligent behavior. He is particularly interested in machine learning, with which computers can learn from experience, and decision-theoretic planning, with which they can reason about their goals and deduce behavioral strategies that maximize their utility. In addition to theoretical work on these topics, he has in recent years also focused on applying them to practical problems in robotics and search engine optimization.

Printed in the United States
by Baker & Taylor Publisher Services